SpringerBriefs in Applied Sciences and Technology

SpringerBriefs present concise summaries of cutting-edge research and practical applications across a wide spectrum of fields. Featuring compact volumes of 50 to 125 pages, the series covers a range of content from professional to academic.

Typical publications can be:

- A timely report of state-of-the art methods
- An introduction to or a manual for the application of mathematical or computer techniques
- A bridge between new research results, as published in journal articles
- A snapshot of a hot or emerging topic
- An in-depth case study
- A presentation of core concepts that students must understand in order to make independent contributions

SpringerBriefs are characterized by fast, global electronic dissemination, standard publishing contracts, standardized manuscript preparation and formatting guidelines, and expedited production schedules.

On the one hand, **SpringerBriefs in Applied Sciences and Technology** are devoted to the publication of fundamentals and applications within the different classical engineering disciplines as well as in interdisciplinary fields that recently emerged between these areas. On the other hand, as the boundary separating fundamental research and applied technology is more and more dissolving, this series is particularly open to trans-disciplinary topics between fundamental science and engineering.

Indexed by EI-Compendex, SCOPUS and Springerlink.

Abimbola Jacob Olasoji · Sang Hyuk Im

Functional Coating Layers for Surface Engineering of Electronic Displays

Springer

Abimbola Jacob Olasoji [iD]
Department of Chemical and Biological
Engineering, BK21 Four R&E Center
Korea University/ERC
Seoul, Korea (Republic of)

Sang Hyuk Im
Department of Chemical and Biological
Engineering, BK21 Four R&E Center
Korea University/ERC
Seoul, Korea (Republic of)

ISSN 2191-530X   ISSN 2191-5318 (electronic)
SpringerBriefs in Applied Sciences and Technology
ISBN 978-3-032-12842-3 ISBN 978-3-032-12843-0 (eBook)
https://doi.org/10.1007/978-3-032-12843-0

This Springer imprint is published by the registered company Springer Nature Switzerland AG
The registered company address is: Gewerbestrasse 11, 6330 Cham, Switzerland

If disposing of this product, please recycle the paper.

To all researchers, engineers, and innovators who continually push the boundaries of optical science to create clearer, brighter, and more adaptive display technologies that shape how we see and experience the world. Your curiosity transforms vision into reality—making the invisible, visible:

May your light continue to illuminate new frontiers of discovery.

Preface I

The evolution of modern display technologies has been nothing short of revolutionary. From the early days of cathode ray tubes to the contemporary landscape of OLED, Micro-LED, and quantum-dot displays, the pursuit of visual perfection continues to redefine how humans interact with information. Yet, beneath every vivid color, sharp contrast, and flexible surface lies a critical but often understated component—the functional coating layer.

Functional coatings represent the convergence of optics, chemistry, and materials science. They not only safeguard display surfaces from mechanical and environmental stress but also engineer light itself—modulating reflection, refraction, polarization, and scattering to deliver the precision visual experience demanded by today's high-performance devices. The emergence of foldable, stretchable, and 360-degree display formats further expands the frontier, compelling the design of coatings that can deform without optical loss, self-heal, or even adapt to new form factors.

As the electronics industry transitions toward sustainability and multifunctionality, surface engineering becomes a central enabler. Advanced coatings—antireflective, anti-glare, anti-smudge, hydrophobic, or self-repairing—now embody both functional necessity and environmental responsibility. The integration of nanostructured materials, metamaterials, and bio-inspired architectures provides unprecedented control over light–matter interaction, paving the way for durable, lightweight, and energy-efficient display systems.

This SpringerBrief, *Functional Coating Layers for Surface Engineering of Electronic Displays*, offers a rigorous yet accessible exposition of these developments. The authors bring together foundational optical theory, coating materials chemistry, process engineering, and emerging industrial practices into a cohesive narrative. Their work highlights how innovations in deposition technologies and interfacial design directly translate into enhanced performance and manufacturability across diverse display applications—from mobile devices to automotive and large-area systems.

By bridging scientific fundamentals with practical implementation, this volume will serve as a valuable reference for students, researchers, and professionals engaged in optoelectronics, materials science, and advanced display manufacturing. It offers both depth of understanding and breadth of vision—qualities essential for guiding the next generation of surface engineering solutions that will shape the visual technologies of tomorrow.

Seoul, South Korea   Prof. Sang Hyuk Im

Preface II

Over the past decade, the display industry has witnessed a profound transformation—from rigid flat panels to ultrathin OLED smartphones, large-area micro-LED billboards, flexible automotive dashboards, and emerging three-dimensional visualization systems. At the core of these innovations lies the subtle yet decisive influence of functional coating layers, which govern optical clarity, energy efficiency, durability, tactile response, and environmental stability.

This book provides researchers, engineers, and graduate students with an integrated resource that extends beyond cataloging materials to elucidate how coatings operate, how they are fabricated, and how their performance can be optimized for a wide spectrum of electronic display applications. Each chapter follows a coherent progression—from human visual perception and optical fundamentals to coating design principles, deposition methodologies, and practical case studies—culminating in emerging directions such as self-healing hybrids, eco-friendly fluorine-free oleophobic coatings, and quantum-dot/OLED light-outcoupling layers.

The objective is to furnish readers with both the theoretical insight and engineering framework required to design, characterize, and scale high-performance optical coatings for the rapidly advancing display landscape. It is hoped that this volume will contribute to the broader understanding and continued evolution of surface-engineered electronic displays.

Seoul, South Korea

Prof. Sang Hyuk Im

Acknowledgments The authors gratefully acknowledge the support and collaboration of colleagues and students in the Nano Energy Convergence System Laboratory at Korea University. Special appreciation is extended to Prof. Sang Hyuk Im for his invaluable mentorship and guidance. We also thank the broader scientific community—whose pioneering work in thin-film optics, materials science, and nanofabrication laid the foundation for the concepts explored in this book.

We appreciate the constructive feedback of reviewers, editors, and peers who have challenged us to refine and strengthen this work. Finally, we recognize the unwavering support of family and friends whose encouragement has sustained this research and writing journey.

Competing Interests The authors have no competing interests to declare that are relevant to the content of this manuscript.

Contents

About the Authors

Abimbola Jacob Olasoji received his B.Sc. in Chemical Engineering from Obafemi Awolowo University, Ile-Ife, Nigeria, in 2013. He won the prestigious Total E&P/NNPC Joint Venture Scholarship, through which he completed an M.Sc. in Petroleum Engineering and Project Development in 2016 at the Institute of Petroleum Studies, University of Port Harcourt, in collaboration with IFP School, France. He further won the highly competitive Global Korea Scholarship (GKS) to pursue his Ph.D. in Chemical Engineering at Korea University, where he is currently conducting doctoral research under Prof. Sang Hyuk Im. His research bridges polymer nanoparticle synthesis, colloidal self-assembly, and interfacial fluid dynamics in porous systems, with emphasis on fabrication of inverse opal thin films for optoelectronic applications, including sensitized and perovskite solar cells, light-emitting diodes, photodetectors, and advanced optical sensors.

Sang Hyuk Im received his Ph.D. in Chemical and Biomolecular Engineering from the Korea Advanced Institute of Science and Technology (KAIST) in 2003. He has held research and academic positions at the University of Washington, LG Chem Research Park, and the Korea Research Institute of Chemical Technology (KRICT). After serving as an Associate Professor at Kyung Hee University, he joined the Department of Chemical and Biological Engineering at Korea University, where he is currently a Professor. His research focuses on material design and application of metal-halide perovskites for solar cells, LEDs, detectors, photobatteries, photocapacitors, quantum sensing, and carbon-neutral technologies.

Chapter 1
Introduction

This chapter introduces the global context of the electronic display industry, which is projected to expand steadily at a CAGR of 5.1% between 2024 and 2029, with revenues increasing from USD 135.2 billion to USD 173.7 billion. The surge is largely attributed to the adoption of OLED and AMOLED technologies in IoT platforms, valued for their low power consumption, superior contrast, and flexible form factors. The narrative highlights the technological trajectory of displays, tracing the shift from cathode ray tubes (CRTs) to liquid crystal displays (LCDs) and, more recently, to active-matrix OLED technologies. These advances have introduced unprecedented form factors, including foldable, ultra-thin, and flexible screens with multitouch and 3D rendering capabilities, reshaping device performance and user interaction.

The discussion then traces the technological aspirations that spurred the development of flat panel displays (FPDs), particularly the push for lightweight, wall-mounted televisions. This pursuit gave rise to a wide spectrum of non-emissive and self-emissive display categories. Non-emissive displays, such as LCDs and electrochromic displays (ECDs), depend on external light sources and optical modulation, while self-emissive systems—including EL, VF, PDP, and LED displays—generate light intrinsically, offering superior contrast, response time, and viewing angles. These distinctions form the backbone of contemporary display innovation.

The chapter concludes by underscoring the indispensable role of functional optical coatings in resolving inherent optical limitations. Beyond television panels, coatings now enable robust performance across immediate consumer, industrial, defense, and biomedical applications. In particular, the advancement of anti-reflection and anti-glare films is identified as a prime mover for the optical clarity, resilience, and durability required in next-generation electronic displays.

A. J. Olasoji and S. H. Im, *Functional Coating Layers for Surface Engineering of Electronic Displays*, SpringerBriefs in Applied Sciences and Technology, https://doi.org/10.1007/978-3-032-12843-0_1

The global electronic display industry is poised for sustained expansion, with market projections indicating a compound annual growth rate (CAGR) of 5.1% over the forecast period from 2024 to 2029. The industry revenues are expected to rise from USD 135.2 billion in 2024 to approximately USD 173.7 billion by 2029 [1]. This growth trajectory is largely driven by the increasing integration of advanced display technologies—particularly organic light-emitting diode (OLED) panels—into Internet of Things (IoT) applications, where low power consumption, high-contrast ratios, and flexible form factors are in high demand. Over recent decades, the display technology landscape has undergone significant transformation, evolving from traditional cathode ray tube (CRT) systems to flat-panel liquid crystal displays (LCDs), and more recently, to sophisticated active-matrix OLED (AMOLED) display technologies. These emerging platforms are reshaping the display paradigm, introducing novel capabilities such as ultra-thin flexible screens, foldable form factors, and multitouch responsiveness enhanced by 3D rendering technologies [2]. This technological progression underscores the increasing need for advanced functional coatings that support mechanical resilience, optical clarity, and environmental stability in next-generation display systems.

Electronic display is an output device which consists of display panels used to convey information in textual, pictorial and/or graphical form. It also helps users to have access to information for navigation, healthcare, education, and product availability at a department store. Furthermore, electronic displays are integral parts of modern restaurants, medical centers, military applications, point of sale systems of various service industries. Overall, electronic displays are indispensable in present industrial revolution era. The evolution of electronic display technology traces its roots to the cathode ray tube (CRT), which was initially observed as cathode rays within an evacuated glass tube by Sir William Crookes in the late nineteenth century. The foundational development into a usable display device is credited to Karl Ferdinand Braun, who in 1897 constructed the first cathode ray oscilloscope, thus marking a pivotal step in display innovation [3]. The CRT functions by emitting a focused beam of electrons from a heated cathode situated in a vacuum-sealed tube. These electrons are directed and accelerated toward a phosphor-coated anode screen by means of electric and magnetic focusing grids. Upon impact, the kinetic energy of the electrons is converted into visible light by the phosphor layer, resulting in image generation. To form a complete image, magnetic deflection coils scan the electron beam horizontally and vertically across the screen in a raster pattern—line by line and frame by frame—at high frequency. This synchronized beam deflection and modulation enables the rendering of complex visual content on a white display surface, forming the basis for early television and monitor technologies [4].

The aspiration to develop a television that could be mounted on a wall like a framed picture—flat, lightweight, and more aesthetically appealing—catalyzed the innovation of various flat-panel display (FPD) technologies, including efforts to engineer "thin-profile" cathode ray tubes (CRTs). This push gave rise to a diverse array of display systems now collectively classified as flat-panel displays. Among these, commercially significant non-CRT FPDs evolved to serve numerous applications far beyond their initial home-entertainment intent. These FPDs are broadly

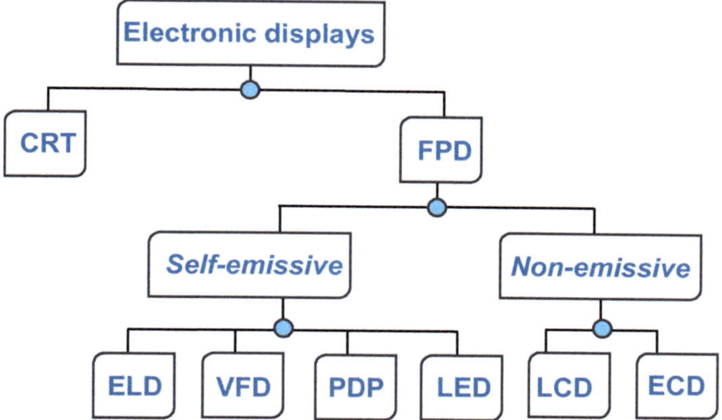

Fig. 1.1 Electronic display technologies: FPD = flat-panel display, CRT = cathode ray tube, ELD = electroluminescent display, VFP = vacuum fluorescent display, PDP = plasma display panel, LED = light-emitting diode, LCD = liquid crystal display, ECD = electrochromic display

categorized into non-emissive displays and self-emissive displays, as illustrated in Fig. 1.1. Non-emissive displays, also referred to as passive displays, do not generate light intrinsically. Instead, they rely on external backlighting and employ optical modulation mechanisms—such as polarization, diffraction, or color filtering—to render visual content. Prominent examples include liquid crystal displays (LCDs) and electrochromic displays (ECDs), which operate by manipulating light from a secondary source. In contrast, self-emissive displays produce light through direct conversion of electrical energy into optical emissions, thereby eliminating the need for backlighting. These include electroluminescent (EL) displays, vacuum fluorescent displays (VFDs), plasma display panels (PDPs), and light-emitting diode (LED) displays. These FDP technologies offer significant advantages in terms of contrast ratio, response time, and viewing angle and have become foundational in modern display design [3].

LCD device was developed by Heilmeier and co-workers after the idea of applying liquid crystal materials for display applications was first conceived in 1963 by Drs. Richard Williams and George Heilmeier at the David Sarnoff Research Center in Princeton, New Jersey [3, 5, 6]. LCDs operate based on electro-optical properties of liquid crystals [7]. Major parts of a typical nematic LCD pixel include a liquid crystal layer, and two transparent electrodes situated between two glass substrates such that the pixel is sandwiched between rear (first) polarizer and front (second) polarizer. A white mode indicates that transmitting axes of the two polarizers are perpendicular while black mode stands for parallel transmitting axes. The pixel is in an "on" state when the second polarizer transmits light in white mode due to no voltage applied to the liquid crystal layer. In contrast, when voltage is applied to liquid crystal pixel, the liquid crystal molecules align themselves with electric field as the polarizer blocks the light to put the pixel in "off" state [4]. Due to the achievement of low-cost and low-power display with quality visibility LCDs, application was extended to use in automobile dashboards, aircraft cockpit displays, telephones, gaming machines, monitoring and control systems in automatic machines, and many portable devices [3].

ECD was commercialized by an Irish start-up known as NTERA. Electrochromic displays operate based by utilizing viologens in some cases, which reversibly change color in aqueous solution when voltage is applied. Also, electrochromic displays are quasi-bistable such that small amounts of energy are only required to preserve their present state but quite a high amount of energy to switch between states. According to NTERA ECD company, features of patented state-of-the-art EC nanotechnology include sunlight readability showcasing great performance in all lighting conditions, contrast ratio of four times that of LC display, cost-effective production, reduced power consumption due to bistability, low-voltage direct current (DC) operation possible, sophisticated lightweight design with less than 2 mm thickness, development work on flexible substrates compatible [8].

EL also called electrophotoluminescence was first discovered in 1936 in Paris by Georges Destriau who observed that light can be emitted when an electric field was applied to an oil suspension made of zinc sulfide containing copper [9]. The eventual development of alternating current (AC) thin-film EL displays was achieved by Sharp Corporation (Nara, Japan) in 1970s by successfully making matrix display with long life and brightness [3]. EL displays are made of layers of metal–insulator electroluminescent layer-insulator-conductors which are all deposited by thin-film techniques on glass plates. EL displays use very low current at high AC voltage (100 V) above a threshold to produce visible light glow (orange-yellow) emitted from central layer as multiplexing is required to drive a dot matrix panel. EL electronic displays can operate at $-$ 40 to 80 °C because they contain no gas unlike LCDs and plasma displays. Considering high improvement made in the display quality, reliability, manufacturing cost reduction coupled with high-voltage, low-current AC power supply requirement, EL panels are now used in aerospace signage, automotive, mobile computers, industrial control systems, and watches [3, 8].

The first VF display was developed to operate at a much lower voltage than CRT. It uniquely offers flat, thin CRT-like display, and it was developed by Dr T. Nakamura Ise Electronics Corporation in 1967 [10]. A typical operating process of VFD involves heating cathode filament to about 600 °C to facilitate emission of thermal electrons. The vacuum fluorescent tubes use ceramic anode substrate which is sealed in a glass bulb. An anode voltage of between 10 and 50 V is applied through this anode as well as grid voltage supplied to grid of selected segment. Subsequently, electrons from the filament accelerate past the grid and anode and collide with phosphor to generate photon emission. As a result of very high reliability of VFDs, they have extensive use in automobiles as primary instruments such as speedometer and fuel gauge and also heads-up displays [3].

PDP (also known as gas-discharge display) was invented by Georges Claude in France in 1915. It was considered to be an outgrowth of neon lamp, and a typical PDP consists of pixels which behave like a small fluorescent light tube. It comprises of a cathode, an anode and a phosphor layer sealed in inert gas. Operating mechanism of PDP involves the application of high-voltage electric field to ionize inert gases such as helium, neon, argon, xenon, or a mixture placed in an airtight sealed glass envelope to produce plasma which radiates ultraviolet light. The phosphors convert the UV radiation into visible RGB light. Furthermore, the electric field that generates excitation can be in the form of either a direct current (DC plasma) type or an alternating current (AC plasma) type. Fabrication of PDP is done by screen-printing process. As a result of this, their pixels cannot be very small. PDP gained ground in the 40–60″ diagonal direct-view display market such as wall-mounted TVs [3, 4].

The development of LED displays was founded on the application of gallium arsenide phosphide (GaAsP) p–n junctions, pioneered by research efforts at institutions such as Bell Laboratories, Hewlett-Packard, RCA Laboratories, and Monsanto Chemical Company. Light-emitting diodes (LEDs) emerged from advancements in semiconductor technology, wherein light is emitted when a forward-biased voltage is applied across a p–n junction formed in a single crystal of group III–V compounds, such as gallium arsenide (GaAs) or gallium arsenide phosphide (GaAsP). The emission of different wavelengths—corresponding to red, green, yellow, and blue light—can be controlled through selective doping or by using ternary and quaternary III–V semiconductor alloys. The first commercial LED displays were introduced by Monsanto and Hewlett-Packard in 1968 [11]. Today, numeric and alphanumeric LED displays are widely integrated into consumer electronics and industrial devices. Additionally, LEDs have increasingly supplanted incandescent lamps as status indicators due to their superior efficiency, compactness, and longevity [3].

OLEDs, often described as the organic counterpart of LEDs, share a similar basic operating principle but differ fundamentally in their light-emission physics. In OLEDs similar to other semiconductor-based light-emitting diodes such as perovskite LED (PeLED) [12], light generation occurs through radiative recombination, where an excited electron recombines with an electron-deficient site, known as a hole, within the emissive organic layer. The cathode facilitates electron injection into the recombination zone, while the anode supplies holes. This process results in efficient electroluminescence without the need for a backlight. However, OLEDs are significantly more sensitive to surface irregularities than LCDs and thus demand extremely smooth substrates to prevent the formation of pinholes, which can severely degrade device performance and shorten operational lifespan. A prominent subset of OLEDs includes polymer light-emitting diodes (PLEDs), which utilize conjugated polymers as the emissive material. For advanced display architectures such as active-matrix OLEDs (AMOLEDs), the integration of an active-matrix backplane is necessary. This involves printing or patterning thin-film transistors (TFTs), capacitors, and the OLED elements themselves to achieve pixel-level control, enabling high-resolution, flexible, and energy-efficient display systems [8].

Flexible displays, also known as flexible flat-panel displays, emerged as a result of rapid advancements in display science and engineering aimed at enhancing traditional flat-panel technologies. Unlike conventional rigid displays, flexible displays utilize alternative substrates such as polymer films or ultra-thin glass, which provide the mechanical pliability required for deformation without compromising electrical or optical performance. These materials enable the development of displays that can bend, roll, fold, or curve while maintaining functionality. The technological transition to flexible substrates has opened new possibilities in device design and user experience. Key advantages include ultra-lightweight construction, thin form factors, and increased mechanical durability, making flexible displays highly suitable for portable electronics, foldable smartphones, wearable devices embedded in textiles, and curved dashboard interfaces in vehicles. Moreover, this flexibility offers engineers greater design freedom, allowing integration of displays into unconventional surfaces and dynamic environments, further pushing the boundaries of innovation in next-generation electronics [13].

Despite the significant advancements chronicled in the evolution of electronic display technologies, the indispensable role of optical coatings within these systems cannot be overstated. Their primary function lies in resolving optical challenges that inherently affect display performance, particularly those related to light management, visibility, and surface interaction. Far beyond their conventional applications in television panels, optical coatings have evolved into multifunctional solutions across a wide range of sectors. These include personal and commercial computing and entertainment devices, industrial systems operating under large-scale or hazardous conditions, military technologies that demand specialized display capabilities, and biomedical or healthcare applications where precision, hygiene, and even antimicrobial surfaces are paramount [2]. Crucially, the sustained advancement of electronic display technologies is intimately tied to excellent parallel progress in the development of material systems used in device fabrication. Thus, enhancing optical coating technologies—specifically through the advancement of anti-reflection (AR), anti-glare (AG), and other functional coatings—are instrumental in shaping the next generation of high-performance, resilient electronic displays. As a foundational step, it is essential to understand how the material properties of optical coatings influence image quality and visual perception. This begins by examining how the human eye interacts with light from electronic displays and how optical coatings modulate that interaction at the surface level to optimize visual output.

References

1. *Display Market Size, Share and Growth Analysis by MarketsandMarkets*™ [Internet]. Available from: https://www.marketsandmarkets.com/Market-Reports/display-market-925. html. Published: July 2024, Report code: SE 3387
2. D. Saddington, *Optical Coatings for Imaging and Displays* (IntertechPira, Maine, USA, 2012)
3. A. Joseph, *Castellano Handbook of Display Technology* (Gulf Professional Publishing, California, USA, 1992)
4. L. Li, *Optical Coatings for Displays* (National Research Council of Canada, Ottawa, Ontario, Canada, 2003)
5. R. Williams, Chem. Phys. **39**, 384 (1963)
6. R. Williams, G.H.J. Heilmeier, Chem. Phys. **44**, 638 (1966)
7. P. Yeh, C. Gu, *Optics of Liquid Crystal Displays. Wiley Series in Pure and Applied Optics* (Wiley, New York, 1999)
8. W. Bock, *Advances in Flexible Electronics Displays* (Intertech-pira, United Kingdom, 2005)

9. G. Destriau, J. Chim. Phys. Phys. Chim. Biol. **33**, 587 (1936)
10. K. Kasano, M. Masuda, T. Shimojo, K. Kiyozumi, Proc. SID **21**(2), 107 (1980)
11. L.E. Tannas, *Flat Panel Displays and CRTs* (Van Nostrand Reinhold Company, New York, 1985), p. 289
12. A.J. Olasoji, J.K. Park, H.J. Lee, Y. Song, D.S. Lee, S.H. Im, Metal halide perovskites: a platform for next-generation multifunctional devices. Adv. Ind. Eng. Chem. **1**(1), 1–69 (2025)
13. P. Gregory, *Crawford Flexible Flat Panel Displays* (Wiley, West Sussex, England, 2005)

Chapter 2
Human Visual Perception and Fundamental Optical Principles Related to Electronic Display Technologies

This chapter establishes the critical link between the human visual system and the performance requirements of electronic displays. The human eye is highly sensitive to luminance, contrast, chromaticity, resolution, and dynamic range, and these parameters directly define the standards for display engineering. Visual perception depends on photoreceptors—cones and rods—that respond to wavelengths in the 400–700 nm visible spectrum, with trichromatic sensitivity enabling color reproduction through red, green, and blue channels. Perceptual thresholds, including spatial resolution (≈ 1 arc minute), response times (~ 0.1 s), and contrast sensitivity, impose strict conditions on how display outputs must be optimized to ensure comfortable, accurate, and fatigue-free viewing.

Alongside visual perception, the chapter outlines the optical principles that govern light–surface interactions in display systems. Concepts such as reflection, refraction, transmission, absorption, and scattering define how light propagates through layered structures. At the display interface, Fresnel reflection at air–substrate boundaries leads to optical losses, glare, and reduced image fidelity, necessitating anti-reflective strategies. Thin-film interference, refractive index matching, and gradient index coatings are discussed as essential mechanisms for minimizing reflection and enhancing transmission. Accordingly, interference-based anti-reflective designs, refractive index matching, and polarization management are introduced as indispensable strategies for ensuring brightness, wide-viewing angles, and stable color reproduction.

To translate theory into engineering practice, design tolerance budgets are presented (Table 2.1). These outline acceptable ranges for coating thickness (± 1–2 nm), refractive index deviation (± 0.01–0.02), absorption ($k < 0.01$ at 550 nm), haze (< 1–2%), color shift ($\Delta E^* \leq 2$), and angular deviation (< 2–$3°$). Together, these metrics define the optical and mechanical precision required in coating deposition processes. By linking the thresholds of human vision with

A. J. Olasoji and S. H. Im, *Functional Coating Layers for Surface Engineering of Electronic Displays*, SpringerBriefs in Applied Sciences and Technology, https://doi.org/10.1007/978-3-032-12843-0_2

the physics of light–surface interaction, this chapter emphasizes the necessity of advanced functional coatings in display technologies. These layers are not passive finishes but optical regulators that suppress reflection, mitigate scattering, and maintain polarization stability. In doing so, they directly shape clarity, comfort, and clarity of the visual output. Acting as the interface between emitted light and human perception, coatings ensure next-generation displays achieve consistent, high-quality performance across diverse environments.

Table 2.1 Representative design tolerance budgets for optical coatings in electronic displays [9–14]

Parameter	Target value/limit	Design consideration	Impact on display performance	Notes
Layer thickness (d)	$\lambda/4n \pm 1$–2 nm	Quarter-wave design precision	Controls destructive interference	Deviation causes $\Delta E^* > 2$ (visible color shift)
Refractive index (n)	$\pm$ 0.01–0.02	Material/process uniformity	Maintains AR condition	Critical for wide-viewing angles
Extinction coefficient (k)	≤ 0.01 (visible)	Low absorption coatings	Brightness retention	Higher k reduces transmission
Residual reflectance (R)	< 1% @ 550 nm	Absorption-related reflectance control	Minimizes brightness loss	550 nm = peak human eye sensitivity
Haze (H)	< 0.5–1%	Surface smoothness/scattering control	Image clarity	Excess haze reduces resolution
Color stability (ΔE^*)	< 2	Angular/processing tolerance	Avoids perceptible color shift	$\Delta E^* > 2$ visible under standard conditions
Angular deviation	< 2–3°	Refractive index control	Maintains color accuracy	Essential for wide-view applications

In the realm of modern display technologies, functional optical coatings serve as critical interfaces between the emitted light from screens and the perceptual capabilities of the human eye. These coatings are engineered not merely for aesthetic or protective purposes, but to actively enhance the visual experience by manipulating how light behaves on the display surface. The human visual system is highly sensitive to brightness, contrast, color uniformity, and glare—all of which can be influenced by how coatings control reflection, refraction, absorption, and scattering of light. Anti-reflective coatings, polarization layers, interference films, and spectral filters are tailored to optimize visual clarity while minimizing eye strain and ambient light distortion. Understanding the physiological parameters of human sight—such as luminance sensitivity, angular resolution, and color perception—is therefore essential in the formulation of advanced coating materials. This chapter explores how functional optical coatings are designed in response to the visual system's requirements and the optical principles that govern light–material interactions in display environments, ultimately driving performance in high-resolution, energy-efficient, and user-comfort-optimized electronic displays.

2.1   The Relationship Between the Human Visual System and Electronic Display Output

In the context of electronic display systems, the human eye functions as the primary optical sensor, and its physiological and perceptual characteristics serve as fundamental benchmarks for display performance optimization. The eye is sensitive exclusively to the visible spectrum of electromagnetic radiation, spanning wavelengths from approximately 400–700 nm, and color perception is mediated through three types of cone photoreceptors—each optimally responsive to red, green, or blue (RGB) light. This trichromatic response forms the basis for RGB color encoding in display technologies. Due to the limited diameter of the eye's pupil (aperture), the spatial resolving capability is constrained to approximately 1 arcminute, defining the smallest visual angle at which two distinct points can be differentiated.

Critical parameters such as color fidelity, luminance, contrast ratio, and temporal response are intrinsically governed by the optical and neural properties of the human visual system. Visual detection also requires a threshold level of luminance, which determines the minimal light intensity perceivable by the human eye under specific ambient conditions. Similarly, luminance contrast—the ratio between the light intensities of adjacent pixels (e.g., "on" vs. "off" states)—is essential for visual distinction between objects and background. Therefore, the essential visual attributes of an electronic display—namely brightness (measured in candela per square meter, cd/m^2), color rendering accuracy, pixel resolution, refresh rate, and contrast ratio—must be engineered to align with the perceptual thresholds and preferences of human vision [1]. Furthermore, the temporal resolution or response time of the human eye is around 0.1 s, influencing the minimum refresh rate required for smooth image transitions and motion portrayal. In essence, the optical design and performance standards of display technologies are intrinsically tied to the sensory capabilities of the eye. These physiological constraints form the foundation upon which display specifications are defined, ensuring that visual output remains legible, comfortable, and perceptually accurate across varying applications and viewing conditions [2].

Ultimately, the performance of electronic displays is intrinsically linked to their interaction with the human visual system, which is sensitive to parameters such as luminance, contrast, color accuracy, spatial resolution, and glare. Functional optical coatings serve as critical mediators in this interface by controlling light behavior—through reflection, transmission, and interference—to optimize visual clarity and comfort. By leveraging thin-film interference principles, these coatings enhance display readability, reduce glare, and preserve color fidelity across diverse lighting conditions. As such, they play a fundamental role in aligning the optical output of display technologies with the perceptual capabilities of the human eye [3–6].

2.1.1   Visual Sensitivity and Display Coating Optimization

Human visual sensitivity follows a nonlinear response, with the eye most responsive to wavelengths around 555 nm (green-yellow range) under photopic (daylight) conditions. Functional coatings—such as spectral filters or interference layers—can be engineered to modulate the spectral power distribution (SPD) of displays to align better with this peak sensitivity, improving both brightness efficiency and perceived image sharpness. Blue-light filtering coatings, for example, are increasingly used to reduce high-energy visible (HEV) light in the 400–450 nm range, which may cause visual discomfort and circadian rhythm disruption.

2.1.2   Contrast Perception and Anti-reflective Coatings

The human eye perceives contrast rather than absolute luminance. Hence, maintaining high-contrast ratios is key to display legibility, especially under ambient lighting conditions. Functional anti-reflective (AR) coatings minimize surface reflections that would otherwise reduce contrast by washing out black levels. These coatings are designed using multilayer interference principles, often with alternating high- and low-index materials (e.g., TiO_2/SiO_2 stacks) to destructively interfere with reflected ambient light and improve readability, especially in mobile or outdoor display applications.

2.1.3   Glare, Ambient Light, and Surface Scatter

Glare and specular reflections interfere with the human eye's ability to resolve fine details and colors, especially on glossy display surfaces. Anti-glare coatings, which are micro- or nanotextured surfaces, diffuse incoming light to reduce reflected luminance peaks. While this introduces mild image blur, it balances visibility and comfort in high-glare environments such as automotive dashboards or public signage. The angle-dependent reflectivity and haze can be tuned during coating design to find an optimal compromise between image clarity and glare suppression.

2.1.4 Viewing Angle Tolerance and Polarization Management

The human field of vision spans nearly 180°, but displays are typically optimized for narrower viewing cones due to brightness falloff and color shifting at oblique angles. Functional coatings that incorporate optical retarders or polarization-preserving films help control light polarization state, allowing for consistent brightness and color uniformity across a wider range of viewing angles. This is particularly important in wide-screen TVs, 3D displays, and automotive or aviation panels where users interact with the display from variable positions.

2.1.5 Color Fidelity and Interference Control

Color perception depends on both the spectral content and uniformity of emitted or reflected light. In high-resolution displays like OLED (organic) or QLED (quantum-dot) panels, interference coatings and color filters help narrow the bandwidth of emitted light to enhance color saturation and distinguishability. Advanced interference coatings can even serve to correct color shifts due to angular viewing, ensuring accurate chromaticity from various directions—essential for professional applications such as medical imaging, design, and broadcasting.

2.1.6 Temporal Sensitivity and Flicker Minimization

Although not a coating issue in isolation, the eye's temporal sensitivity to flicker—particularly at lower refresh rates—is affected by optical layers that influence luminance modulation. Some emerging optical coatings are being designed with photoresponsive or electrochromic properties that can respond dynamically to ambient conditions, potentially reducing flicker artifacts in adaptive brightness systems.

The human visual system is not a passive observer of light but a finely tuned biological sensor that demands precision in luminance, contrast, and spectral output. Functional optical coatings must therefore be developed not only with optical physics in mind but also with an understanding of visual ergonomics. Coatings that minimize glare, enhance contrast, maintain chromatic accuracy, and control light at the micro- and nanoscale are critical to delivering the immersive, comfortable, and accurate viewing experience that modern display users expect.

2.2 Fundamental Optical Principles Governing Light Interaction with Display Surfaces

The behavior of light as it interacts with the surface of an electronic display defines not only the visual performance of the device but also dictates the selection and engineering of functional optical coatings. These coatings are not passive layers; they are precision-engineered films designed to modulate light through reflection, transmission, refraction, absorption, scattering, and interference. To effectively design and implement coatings that improve visual clarity, contrast, color fidelity, and energy efficiency, it is essential to understand the fundamental optical principles that govern light–matter interactions at the interface of display surfaces.

Human eyes become strained as a result of optical disturbance due to light reflection from a display screen when there is change in refractive index (RI) as light travels from one medium through another. Hence, it is important to consider the wave properties of light during propagation and interaction with surfaces of materials for optimal design of display outputs. Depending on the nature of a material, incident light could be absorbed, transmitted (refracted), and reflected by it. Generally, films coated on display surface set and determine how light interacts with the screen.

A comprehensive understanding of optical principles—ranging from light scattering to Fresnel reflections, to thin-film interference, and polarization—is foundational to the development of high-performance display coatings. Functional optical coatings serve not only as protective barriers but as precision-engineered optical components that shape the light reaching the human eye. By manipulating how light interacts with surface materials at micro- and nanoscale levels, these coatings enable higher brightness, better contrast, wider viewing angles, and lower power consumption in today's (conventional)—and tomorrow's (emerging and future)—electronic displays.

2.2.1 Scattering and Surface Roughness

Scattering occurs when light is deflected by surface irregularities or embedded particles, leading to diffusion of light that can enhance or degrade display performance. Controlled scattering, as used in anti-glare coatings, can reduce mirror-like reflections while preserving sufficient image clarity. However, unintentional scattering due to surface roughness or particulate contamination increases haze and reduces display contrast. Light scattering is said to take place when light collides with particles or small surface features based on different scattering models. Geometric scattering happens when particles are significantly larger than the wavelength of light while Mie scattering (named after Gustav Mie) happens when particles are about comparable size as the wavelength of light. Rayleigh scattering (named after John Rayleigh) takes place when the particles are smaller than the wavelength. An example of Rayleigh scattering which mainly takes place in a gas volume is blue coloration of the sky. These effects are wavelength dependent; hence, there exists particle-size-to-wavelength relation [7].

The scattering coefficient (S) is typically a function of the root-mean-square (RMS) roughness σ, and optical performance must be evaluated using techniques like angular-resolved scattering (ARS) or haze measurement according to ASTM D1003. To address this, nanostructured coatings with subwavelength roughness profiles—such as moth-eye structures [8]—are increasingly employed to minimize reflectance while preserving transparency and minimizing scatter.

2.2.2 Interference and Spectral Selectivity

Interference phenomena are crucial in multilayer coatings where constructive and destructive wave interactions determine the spectral properties of reflected or transmitted light. Thin-film interference enables the design of narrow-bandpass filters, cold mirrors, and low-pass coatings that selectively control which wavelengths are permitted to pass through or reflect from the display surface. For example, dielectric mirrors in heads-up displays (HUDs) or augmented reality systems utilize Bragg stacks, which consist of alternating high- and low-index layers with thicknesses engineered for constructive reflection at desired wavelengths. Similarly, optical coatings on OLED panels often include microcavity structures formed by interference layers to enhance specific color emissions and improve device efficiency.

2.2.3 Refraction and Viewing Angle Considerations

Light refraction simply means the transmission of light through a material medium other than air. Based on the principle of refraction, the propagation characteristics of light as an electromagnetic wave depend on the material it travels through. Light refraction is said to take place due to speed difference between two media with different optical densities (refractive indices) which causes a change in course of direction when light passes their border. The amount of refraction also depends on wavelength as white light splits into spectral colors [7]. Refractive index design principle is employed for application of anti-reflection coatings on display screens.

Refraction, or the bending of light at an interface, is described by Snell's law:

$$n_1 \sin \theta_1 = n_2 \sin \theta_2 \tag{2.1}$$

where θ_1, θ_2 are the angles of incidence and refraction, and n_1, n_2 are the refractive indices of the respective media.

Functional coatings with gradient refractive indices or birefringent properties (e.g., optical retardation films) are used to manage these effects by aligning phase fronts and optimizing the angular color performance, especially in wide-viewing-angle applications such as televisions and automotive displays.

In displays, unwanted refraction can lead to angular-dependent color shifts and brightness loss, especially in high-index layered structures. Even small refractive index mismatches of $\pm$ 0.02 between layers can introduce $\Delta E^* > 2$, a perceptible color shift under standard D65 illumination. Gradient-index (GRIN) coatings or birefringent retardation films are engineered to minimize these effects, ensuring wide-viewing-angle stability in televisions, monitors, and automotive displays. To maintain color accuracy, angular deviation of refracted beams must be kept below ~ 2–3°, requiring precise refractive index control during deposition. Table 2.1 depicts representative design tolerance budgets for optical coatings in electronic displays, with all the parameters (thickness, n, k, reflectance, haze, ΔE^*, angular deviation), their target values/limits, design considerations, impacts on performance, and notes [9–14].

2.2.4 Absorption and Energy Management

Optical absorption is an essential factor in light modulation and energy dissipation within display stacks. While excessive absorption leads to loss of brightness and heat buildup, strategic absorption can be used for functionality—such as neutral density filters or black matrix coatings used to enhance pixel contrast. The absorption of a material is characterized by the extinction coefficient k, and the absorption depth δ is given by:

$$\delta = \frac{\lambda}{4\pi k} \tag{2.2}$$

Functional coatings with precisely tuned k-values are used to control how deeply light penetrates into a material before being attenuated. In some reflective displays (e.g., e-paper or transflective LCDs), absorption is modulated to optimize ambient light readability and improve power efficiency.

For display coatings, excessive absorption reduces brightness and generates thermal load. To preserve > 95% transmission, the extinction coefficient across the visible spectrum must be $k < 1 \times 10^{-3}$. If k rises above 10^{-2}, transmittance losses exceed 1%, leading to gray or yellow tinting. In OLED or QD-based devices, even tighter thresholds ($k < 5 \times 10^{-4}$) are imposed to suppress photobleaching. Design tolerance budgets typically limit residual reflectance due to absorption losses to less than 1% at 550 nm, the spectral sensitivity peak of the human eye (Table 2.1).

2.2.5 Polarization Effects in Display Coatings

Polarization describes the orientation of the electric field vector of light and plays a crucial role in display systems that rely on polarization states—such as LCDs, 3D stereoscopic displays, or optical sensors. Coatings such as linear polarizers, circular polarizers, and wave plates are applied to manage and preserve polarization states. For instance, in LCD displays, crossed polarizers and compensation films are used to manipulate light transmission depending on liquid crystal alignment. Functional coatings with anisotropic refractive indices are often engineered using polymeric liquid crystals or stretched polymer films to provide phase retardation and reduce angular color shifts.

Here, anisotropic coatings with birefringence Δn in the range of 0.05–0.2 are introduced. To maintain contrast ratios > 1000:1, retardation films must stay within $\pm$ 2 nm phase thickness error; otherwise angular color shifts exceeding $\Delta E^* = 2$–3 occur. Polymeric liquid–crystal coatings and stretched films are engineered with precise molecular alignment to achieve stable retardation, ensuring wide-viewing angles without visible color distortion (Table 2.1).

2.2.6 Reflection and Transmission: Managing Optical Interfaces

When light encounters the boundary between two materials with different refractive indices, a portion is reflected, and the remainder is transmitted. The magnitude of reflectance, R and transmittance, T at normal incidence is governed by the Fresnel equations. Without coatings, the air–glass interface in a typical display reflects approximately 4–8% of incident light per surface, significantly reducing display brightness and contrast. Anti-reflection (AR) coatings are designed by layering materials with alternating refractive indices to produce destructive interference of reflected waves, minimizing total reflectance.

Light reflection takes place due to transition from one medium in which light was initially traveling to another medium such as glass, air, and water. As it bounces off the interface between the two media and changes direction. Optically, these media are characterized by refractive indices, RI which quantifies the speed of light in the current medium compared to speed in a vacuum. Usually, light reflection can be broadly grouped into specular reflection or diffuse reflection. For a flat surface, direct reflection takes place such that both the angle of incidence and that of reflection ray are the same. On the contrary, diffuse reflection takes place due to unevenness of a surface such that the reflected ray of light is not same amount and angle as that of incident ray since the reflected ray of light is dispersed based on the unevenness of the surface [9]. Basic preliminary mathematical model of reflection and refraction for a non-absorbing surface is well put by Fresnel equation which indicates that reflectance, R

measures the amount (fraction) of incident light reflected at the interface while the rest refracted (transmitted) is measured by transmittance, T [10, 15].

Optical performance is enhanced with anti-reflection coatings by adopting interference effects to minimize reflection such that (1) when there is a coating between air and glass substrate, a fraction of incident light reflects at the air/film interface, and some light transmits through and reflects at film/glass interface which travels twice through the film thickness; (2) the optical film thickness is adjusted to ¼ of reference wavelength in optical medium; and (3) then, the reflected light from the film/glass interface is offset by half of wavelength from the reflected light at the air/film interface which interfered destructively to cancel out reflection rays.

Mathematically, model by Fresnel gives basic understanding of light reflection for optimal anti-reflection. The strategy considers a thin film with $RI = n$ on glass substrate $RI = n_s$ as shown in Figure 2.1 with assumption that each interface reflects one wave and the reflected waves have same intensity, and light absorption, while scattering and other optical interactions are negligible. According to Figure 2.1a, the two reflected waves known as R_1 and R_2 can be easily annulled to a no-reflection case if they are made to undergo destructive interference.

π radians out of phase or the phase difference, δ is $n\,\pi/2$, and the thickness of the film (d) is an odd multiple of $\pi/4$ where π is the wavelength of the incident beam. Two essential criteria for anti-reflection are that the reflected waves be π radians (or any odd multiple of π) out of phase (with phase difference, δ) and the film thickness (d) be an odd multiple of $\lambda/(4\,n)$ where λ is the wavelength of the incident beam, ensuring destructive interference at the design wavelength [10, 15].

The reflectance at normal incidence is given by

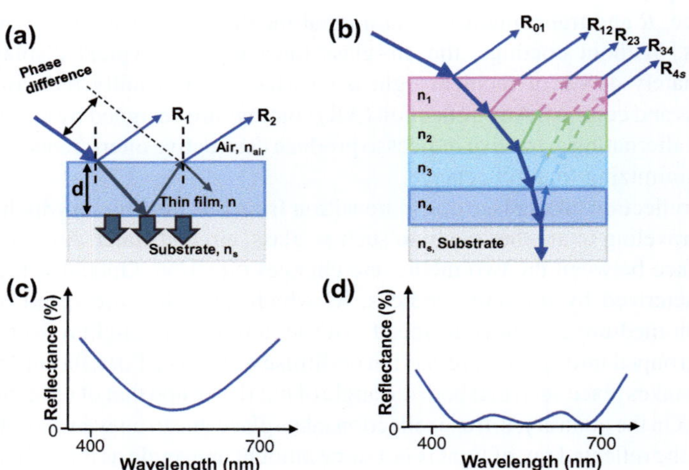

Fig. 2.1 Propagation of light rays through **a** low reflection single-layer film on substrate ($n_s > n$), **b** multilayer film on substrate, **c** reflectance spectra of low reflection (LR) single-layer film, **d** reflectance spectra of anti-reflection (AR) multiple-layer film [15]

$$R = \left[\frac{n_{air}n_s - n^2}{n_{air}n_s + n^2}\right]^2 \tag{2.3}$$

where n_s = refractive index of the substrate, n_{air} = refractive index of air, n = refractive index of the film.

In practice, the refractive index of the coating must be tuned to $n = \sqrt{n_{air}n_s}$. A deviation of $\pm\,0.02$ in n can increase reflectance by 0.5–1%. To suppress reflections below 1% at 550 nm, multilayer AR stacks are fabricated, where each additional layer reduces spectral reflectance but requires thickness control within $\pm\,$1–2 nm. Any deviation beyond $\pm\,$5 nm across a multilayer stack leads to angular-dependent spectral shifts and color instability ($\Delta E^* > 2$). Here, ΔE^* refers to the CIE 1976 perceptual color difference metric, which quantifies perceptible deviations in color between a reference and a measured sample; by convention: $\Delta E^* < 1 \rightarrow$ imperceptible to the human eye $\Delta E^* \sim$ 1–2 $\rightarrow$ perceptible only to a trained observer, $\Delta E^* > 2$ $\rightarrow$ clearly perceptible generally visible to the human eye under standard viewing, $\Delta E^* > 5 \rightarrow$ large, noticeable difference. Haze adds another dimension: to maintain sharpness in handheld displays, haze must remain $< 0.5\%$; in large-area displays up to 1% haze may be acceptable (Table 2.1). This requires particle sizes < 100 nm to avoid Mie scattering.

Since the reflectance proposed by Fresnel is dependent on the s- and p-polarization of light, anti-reflective property is analyzed based on these parameters. Following the basis of single layer on substrate, models for multilayered anti-reflection coating (ARC) can be solved through vector analysis of individual reflected rays such that reflected rays from interface ij between the adjacent layers i and j are given by

$$R_{ij} = |R_{ij}| \exp\left[-2(\delta_i + \delta_j)\right].$$

where $|R_{ij}| = [(n_i - n_j)/(n_i + n_j)]$ and phase thickness of each layer is given by $\delta_i = 2\pi n_i \cos\theta_i d_i / \lambda$ (θ_i is the angle of refraction, d_i is the physical thickness of the layer, and λ is the wavelength of light). Figure 2.1b shows the reflection vectors originating from each of the ij interface, such that the resultant reflection vector is

$$R_{sum} = R_{01} + R_{12} + R_{23} + R_{34} + R_{4s}.$$

where $R_{01} = |R_{01}|$.

$$R_{12} = |R_{12}| \exp[-2(\delta_1)].$$

$$R_{23} = |R_{23}| \exp[-2(\delta_1 + \delta_2)].$$

$$R_{34} = |R_{34}| \exp[-2(\delta_1 + \delta_2 + \delta_3)].$$

$$R_{4s} = |R_{4s}| \exp[-2(\delta_1 + \delta_2 + \delta_3 + \delta_4)].$$

By drastically reducing R_{sum} through adjusting RI and film thickness of each layer, a state of anti-reflection can be attained for display surfaces [10, 15]. Hence, anti-reflective effect is mainly attributed to optical properties such as average and minimum reflectance and reflective wavelength pattern together with factors affecting these optical properties such as the differential of the refractive indices of the layers, the number of the coated thin films, and the thickness of coated layers. The higher the refractive index between the AR thin-film layers, the lower the reflectivity, as the pattern of the reflective wavelength shifts with the thickness of the thin film. Also, the higher the number of thin films, the lower the magnitude of reflection. However, putting the reality of limitations of process and materials, AR is designed to show the lowest reflectivity in the visible light region sensitive to human optic nerve at 550 nm (green) [9]. Figure 2.1c, d show typical reflectance spectrum of low reflection (LR) single-layer film and anti-reflection (AR) multiple-layer film, respectively. The graphs depict that application of LR is only adaptive for the lowest reduction of reflection at about 530–540 nm wavelength (green) while AR incorporated with multiple layers is capable of excellent reduction of reflection across the whole visible light spectrum. This principle is applied in multilayer coatings using materials like MgF_2, SiO_2, and TiO_2 to achieve broadband AR performance in visible displays.

2.2.7 Coating Design Parameters: Thickness, Refractive Index, and Layer Architecture

Finally, the optical performance of any coating system can be reported to be highly sensitive to layer thickness, refractive index, and deposition uniformity. Modern design tools use matrix methods or transfer matrix formalism to model the spectral response of multilayer stacks. In display manufacturing, precision control at the nanometer scale is required during processes such as physical vapor deposition (PVD), atomic layer deposition (ALD), or inkjet printing.

Layer thickness d for interference coatings is typically designed around multiples of the quarter wavelength:

$$d = \frac{\lambda}{4n} \tag{2.4}$$

where d is film thickness, λ is the target wavelength, and n is the refractive index.

For AR coatings optimized at 550 nm, a SiO_2 layer ($n = 1.45$) must be ~ 95 nm. A thickness deviation of $\pm$ 1–2 nm shifts the minimum-reflectance wavelength by ~ 5–10 nm, producing visible green–blue color shifts. For broadband AR stacks, cumulative thickness tolerance across four layers is typically $\pm$ **5 nm** to ensure $\Delta E^* < 2$. Advanced deposition methods (ALD, PVD, or ion-assisted evaporation) must achieve sub-nanometer precision, monitored by in situ ellipsometry or quartz crystal microbalances. Thus, the practical design limits derived from these equations translate directly into optical budgets: thickness $\pm$ 1–2 nm, refractive index

$\pm$ 0.01–0.02, extinction coefficient k < 1 $\times$ 10^{-3}, haze < 0.5%, and color shift ΔE^* < 2. Meeting these constraints is essential for high-resolution, wide-viewing-angle, and color-accurate display performance. Advanced metrology, including ellipsometry and spectrophotometry, is employed to monitor deposition accuracy and ensure conformity with design tolerances.

References

1. G. Wyszecki, W.S. Stiles, *Color Science: Concepts and Methods, Quantitative Data and Formulae* (Wiley, Toronto, 1982)
2. L. Li, *Optical Coatings for Displays* (National Research Council of Canada, Ottawa, Ontario, Canada, 2003)
3. J.A. Dobrowolski, Optical properties of films and coatings, in Handbook of Optics, ed. by M. Bass, E. Van Stryland, D.R. Williams and W.L. Wolfe (McGraw-Hill, New York, 1995), pp. 42.1–42.109
4. A. Thelen, *Design of Optical Interference Coatings. McGraw-Hill Optical and Electrooptical Engineering Series* (McGraw-Hill, New York, 1989)
5. T.C. Schmidt, *Projection Displays, SID Seminar Lecture Notes* (1999), pp. F-6
6. P.A. Keller, *Electronic Display Measurement Concepts, Techniques and Instrumentation* (1997)
7. R.R. Hainich, O. Bimber, *Displays: Fundamentals & Applications* (AK Peters/CRC Press, New York, USA, 2016)
8. A. Gombert, W. Glaubitt, K. Rose, J. Dreibholz, B. Bläsi, A. Heinzel, D. Sporn, W. Döll, V. Wittwer, Subwavelength-structured antireflective surfaces on glass. Thin Solid Films **351**(1–2), 73–78 (1999)
9. Y.J. Hong, Organic/inorganic hybrid coatings for flat panel display. Polym. Sci. Technol. **17**(2), 217–225 (2006)
10. H.A. Macleod, *Thin-Film Optical Filters*, 4th edn. (CRC Press, 2010)
11. M. Born, E. Wolf, *Principles of Optics: Electromagnetic Theory of Propagation, Interference and Diffraction of Light* (Elsevier, 2013)
12. M.S. Shur, R. Zukauskas, Solid-state lighting: toward superior illumination. Proc. IEEE **93**(10), 1691–1703 (2005)
13. J. Rombaut, M. Fernandez, P. Mazumder, V. Pruneri, Nanostructured hybrid-material transparent surface with antireflection properties and a facile fabrication process. ACS Omega **4**(22), 19840–19846 (2019)
14. M.D. Fairchild, *Color Appearance Models*, 3rd edn. (Wiley, 2013)
15. H.K. Raut, V.A. Ganesh, A.S. Nair, S. Ramakrishna, Anti-reflective coatings: a critical, in-depth review. Energy Environ. Sci. **4**(10), 3779–3804 (2011)

Chapter 3
Functional Coatings for Electronic Displays

This chapter explores how functional coatings serve as active interfaces in modern display systems, transforming surface engineering into a critical determinant of performance. As displays evolve toward higher pixel densities, thinner and flexible profiles, and more interactive designs, coatings provide tailored enhancements in optical quality, durability, tactile feel, and environmental stability without altering underlying electronic structures.

The discussion begins with deposition techniques—ranging from sol–gel methods, chemical vapor deposition (CVD), physical vapor deposition (PVD), atomic layer deposition (ALD), and roll-to-roll & printing techniques to advanced approaches such as magnetron sputtering and plasma-assisted methods—that enable the precise fabrication of multilayer thin films. Characterization tools, including spectroscopic ellipsometry, spectrophotometry, and so on are highlighted for assessing thickness, refractive index, surface roughness, and optical performance, ensuring reproducibility and reliability.

Subsequent sections detail the stringent requirements of anti-reflective (AR) films, which must achieve low reflectance, broad spectral coverage, high transmission, and mechanical robustness. The chapter differentiates between AR and anti-glare (AG) coatings: AG relies on controlled surface textures or microparticles to scatter light and reduce glare, while AR employs interference-based multilayer designs with refractive index contrasts, often combining high-index oxides such as TiO_2 or Ta_2O_5 with low-index materials like SiO_2 or MgF_2. Manufacturing strategies, from single- to multilayer stacks and nanoparticle-based composites, are reviewed with attention to how thickness optimization and material selection suppress reflection and preserve image fidelity.

To consolidate these insights, benchmarking tables (Tables 3.1, 3.2) provide side-by-side evaluations of conventional planar LR coatings, moth-eye nanostructures, nanoporous and hollow silica films, and hybrid AR/AG systems.

© The Author(s), under exclusive license to Springer Nature Switzerland AG 2026
A. J. Olasoji and S. H. Im, *Functional Coating Layers for Surface Engineering of Electronic Displays*, SpringerBriefs in Applied Sciences and Technology,
https://doi.org/10.1007/978-3-032-12843-0_3

These comparisons present quantitative optical metrics—reflectance, transmittance, haze—together with durability, angular stability, and cost considerations. By situating performance alongside process feasibility, the benchmarking framework offers a practical roadmap for aligning material choice with application demands, illustrating how functional coatings act as precision-engineered interfaces that elevate display performance and reliability across emerging device platforms.

Table 3.1 Representative process metrics and scalability of functional coating deposition techniques for electronic displays [11–20]

Technique	Typical thickness range	Deposition rate	Substrate scalability	Uniformity/ tolerance	Industrial Adoption
Sol–Gel; dip/ spin/ spray, CBD, electrodeposition	50–500 nm per cycle; up to several μm	10–200 nm/ min (depending on drying/ annealing)	Meter-scale sheets; scalable to roll-to-roll flexible films	± 5–10 nm thickness control	Used in AR/ AG coatings on glass and polymer substrates; low-cost, large-area
PVD; sputtering, evaporation, IBS, HiPIMS	5 nm–5 μm	0.1–1 μm/h (sputtering); faster for evaporation	Gen 8.5–10.5 glass (>3000 × 3000 mm); also flexible web	± 1–2 nm over large areas	Industry standard for ITO electrodes, hard coatings, AR stacks
CVD; APCVD, LPCVD, PECVD	10 nm–several μm	10–100 nm/ min (APCVD/ PECVD); LPCVD slower	Gen 6–10 glass sheets; conformal on 3D substrates	± 2–5 nm, highly conformal	Widely used for SiN_x/SiO_2 barrier layers in OLEDs/ QD displays
ALD	1–100 nm; atomic precision	0.1–5 nm/ min	Flexible films; limited large-area scale, batch, or spatial ALD up to meter rolls	± 0.1–0.3 nm per cycle	Advanced encapsulation, QD/OLED barriers, high-k layers; cost-sensitive
Roll-to-Roll/ Printing; inkjet, slot-die, spray	50 nm–few μm	High throughput; 1–10 m/min web speed	Continuous flexible webs > 1 m wide	± 5–20 nm	Emerging for flexible/ wearable coatings; UV-curable, AR/ oleophobic films

Table 3.2 Standard requirements for anti-reflective films [23]

Entry	SPEC. (AG)	SPEC (Clear LR)	Remarks
Total transmittance (%)	> 92.0	> 95.0	Total amount of light captured by the detector for the amount of incident light, JIS K 7361–1
Haze (%)	5 ~ 30	< 0.5	ASTM D 1003
Reflectance (%)	–	< 3.0	
Gloss(%)	Selection 20 ~ 100	–	The amount of light detected at the 60° reflection angle by irradiating the light at a 60° angle of incidence
Image clarity (%)	> 40	–	HEK 7105
Pencil hardness	> 3H	> H	ASTM D 3363 (based on 500 g standard)
Adhesion	5B	5B	Crosscut Test, ASTM D 3359
Scratch	Good	Good	Steel Wool #0000 test (based on 200~1000 g)

As electronic display technologies continue to evolve toward higher pixel densities, thinner form factors, flexible geometries, and increased user interactivity, the importance of surface-level modifications through functional coatings becomes more pronounced. These coatings are no longer passive protective layers; rather, they are active, engineered interfaces that tailor the optical, mechanical, and environmental performance of the display systems. Surface engineering via functional coatings enables displays to achieve superior visual quality, mechanical durability, tactile response, and environmental robustness without altering the underlying electronics.

Functional coatings have become an indispensable component in the design and engineering of modern electronic display systems. As display technologies evolve to deliver higher luminance, finer resolution, broader color gamut, and enhanced environmental resilience, the role of surface coatings extends far beyond passive protection. These engineered thin films serve critical optical, mechanical, and chemical functions—regulating light transmission and reflection, suppressing glare, enhancing contrast, resisting smudges, and protecting against scratches, moisture, and chemical attack. Whether deployed on rigid LCD panels, flexible OLED substrates, or novel micro-LED and quantum-dot displays, functional coatings are essential to preserving image fidelity, ensuring user comfort, and extending device lifespan. Their application becomes even more critical in portable and touch-enabled devices, where frequent mechanical contact and exposure to ambient light can significantly impair optical performance. This chapter introduces the foundational principles, materials, and multilayer structures behind these coatings, emphasizing their active role in shaping the visual and functional output of advanced electronic displays.

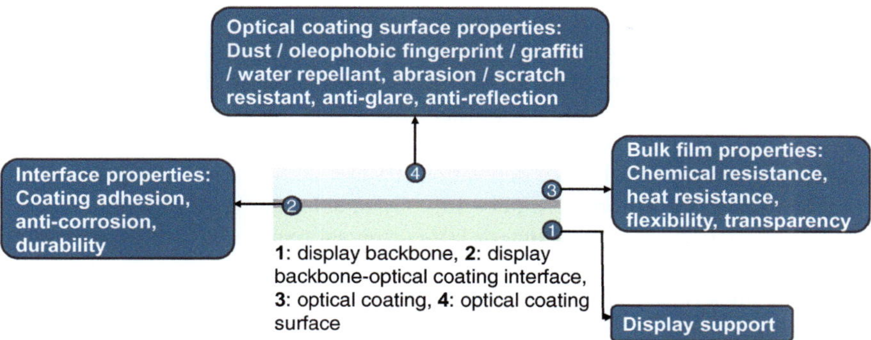

Fig. 3.1 Features of functional coating system for electronic display

Coating is usually applied on electronic surfaces for protective, or specific functional purposes, but in most cases it is a combination of both. Hence, functional coating refers to addition of layer(s) with a particular benefit targeted at satisfying certain demands. A typical functional display coating system is made up of display backbone, display backbone/optical coating interface, optical coating, optical coating/air interface as shown in Fig. 3.1. A functional coating system for electronic displays is an integrated multilayer architecture that enhances display performance through four key feature domains. First, the display support layer forms the structural backbone, ensuring mechanical stability. Next, interface properties between the substrate and coating—such as adhesion, anti-corrosion, and durability—are critical for coating longevity and reliability. The bulk film properties of the coating itself provide essential characteristics like chemical and heat resistance, flexibility, and optical transparency, enabling use across rigid and flexible formats. Finally, the optical coating surface properties ensure user-facing performance by delivering anti-reflection and anti-glare effects, abrasion and scratch resistance, as well as protection against fingerprints, dust, water, and graffiti. Together, these interdependent layers enable modern electronic displays to achieve high image clarity, robustness, and long-term usability under various operational and environmental conditions.

The display support layer forms the structural backbone of the entire device stack, providing mechanical strength, dimensional stability, and surface quality required for reliable optical performance. In conventional rigid panels, this layer is usually an aluminosilicate or soda–lime glass substrate, valued for its high hardness (typically $\geq$ 6–7 H on the Mohs scale), thermal stability well above 500 °C, and ultra-low-surface roughness below 1 nm RMS. These characteristics allow subsequent optical layers to be deposited with nanometer-scale precision and without stress-induced cracking. In flexible and foldable devices, however, the support layer must also tolerate repeated bending and twisting where advanced polymers such as colorless polyimide (PI) and polyethylene naphthalate (PEN) are widely adopted, along with ultra-thin flexible glass (< 100 μm) for premium devices.. These substrates offer low coefficient of thermal expansion (CTE $\approx$ 2–10 ppm/K for flexible glass and 20–40 ppm/K for high-performance PI), good optical transparency ($\approx$ 88–92% transmittance in the visible spectrum), and the ability to withstand > 200,000 bending cycles at $\leq$ 5 mm radius without cracking but are limited by brittleness and handling challenges. Their limited thermal stability (PI ~ 500 °C; PEN ~ 200 °C) constrains deposition processes to low-temperature physical vapor deposition (PVD < 200–250 °C), plasma-enhanced chemical vapor deposition (PECVD ~ 150 °C), or solution-based techniques such as roll-to-roll coating, inkjet printing, or UV-curing approaches to avoid damaging the substrate. The mechanical modulus and surface flatness of this layer directly determine how thick or stiff the subsequent functional coatings can be without inducing residual stress, curling, or delamination.

Equally critical are the interface properties between the substrate and the functional coatings which are primal to long-term reliability. Strong adhesion ensures that optical and protective layers remain firmly bonded during thermal cycling, flexing, or impact. Tailored adhesion-promoting interlayers—such as silane coupling agents, organosilicon primers, or oxygen plasma surface treatments—are commonly used to improve bonding, especially on low-surface-energy polymer substrates. The interface must also provide corrosion and moisture barriers to prevent ingress of water or oxygen that can degrade both the substrate and the functional coating. Additionally, interfacial toughness and resistance to delamination are vital under mechanical stress, particularly in foldable or stretchable devices where cyclic bending can cause crack initiation at poorly adhered interfaces. These engineered interfaces directly influence overall coating durability, scratch resistance, and environmental stability, making them a pivotal design domain alongside the optical functionality of the bulk coating. The adhesion, chemical resistance, and barrier performance of this junction largely determine long-term reliability, especially when displays are subjected to temperature cycling (-40 °C to $+$ 85 °C), moisture ingress, or repeated mechanical strain. Achieving strong adhesion typically begins with surface energy modification. O_2, N_2, or Ar plasma activation can raise polymer surface energy above 60 mN/m, improving wetting and film uniformity. Silane coupling agents such as aminopropyltrimethoxysilane (APTMS) and glycidyloxypropyltrimethoxysilane (GPTMS) create robust Si–O–Si bonds with glass or oxide layers while presenting organic functional groups compatible with polymer or hybrid coatings. UV–ozone treatment is another widely used step to remove low-molecular-weight contaminants and introduce polar groups, often improving adhesive strength by a factor of two to three compared with untreated polymer surfaces. For moisture-sensitive optical films, ultra-thin barrier interlayers deposited by atomic layer deposition (ALD)—such as Al_2O_3 or SiN_x—can lower the water vapor transmission rate to below 10^{-6} g m^{-2} day^{-1}, protecting layers like quantum dots or perovskites. Adhesion and barrier effectiveness are validated using standardized methods including cross-hatch tape tests (ISO 2409, ASTM D3359), peel strength measurements (with > 1–2 N cm^{-1} generally desired for flexible stacks), and scratch resistance testing by pencil hardness ($\geq$ 3H) or nanoindentation (critical load > 1 N before delamination). Well-engineered interfaces ensure that coatings remain intact after tens to hundreds of thousands of folding cycles and resist degradation from sweat, skin oils, and household cleaning chemicals—conditions increasingly relevant to portable and wearable devices.

The bulk film properties of the coating itself define its fundamental optical and mechanical performance across rigid, flexible, and hybrid display formats. This layer must combine chemical and thermal stability with optical clarity and mechanical compliance. Chemically, coatings must resist degradation from humidity, sweat, skin oils, household cleaning agents, and UV light exposure, all of which can cause yellowing, cracking, or delamination. Inorganic oxide films such as SiO_2, Al_2O_3, or TiO_2 deposited by PVD or ALD offer excellent hardness and chemical inertness but are intrinsically brittle, making them prone to cracking under bending or impact. To overcome this, hybrid organic–inorganic nanocomposites—where ceramic nanoparticles (e.g., silica, zirconia) are dispersed in polymeric matrices—are increasingly adopted. These hybrids provide refractive indices tunable between ~ 1.35 and 1.8 while maintaining > 90–95% visible transmittance and elongation at break exceeding 10–30%, far higher than purely inorganic films. Thermally, the bulk coating must survive deposition and device operating temperatures: while glass-based systems may tolerate > 400 °C, flexible polyimide substrates typically require low-temperature coating processes (< 250 °C) to avoid shrinkage and warping. Mechanical flexibility is equally crucial; coatings for foldable displays are routinely subjected to > 100 000 bending cycles at radii below 5 mm. Polymer-toughened oxide nanocomposites or ultra-thin ALD oxides (< 20 nm) on soft underlayers have demonstrated the ability to endure these stresses while retaining barrier properties (WVTR $\leq 10^{-5}$ g m^{-2} day^{-1}). Precise thickness control is fundamental: optical interference layers must remain within $\pm$ 1–2 nm of target design thickness to avoid color shifts ($\Delta E^* > 2$), while index uniformity ($\Delta n \leq \pm 0.02$) ensures angular color stability and consistent luminance. These bulk-layer design parameters are validated by nanoindentation, thermomechanical analysis, and accelerated environmental testing.

Finally, the surface functional attributes of the topmost layer govern the user's tactile experience and the display's resistance to everyday wear. This surface must maintain high optical clarity while providing protection against fingerprints, smudges, scratches, and environmental contaminants. Modern oleophobic and hydrophobic coatings achieve water contact angles > 105° and oil contact angles > 70°, preventing smearing and easing cleaning. Traditionally, fluorinated chemistries such as perfluoropolyether (PFPE) silanes or fluoropolymers were the dominant choice, but regulatory and environmental concerns are driving a shift to fluorine-free systems based on siloxanes, hydrocarbon waxes, and bio-inspired lipid-like structures. These next-generation coatings aim to match the low-surface energy (< 20 mN/m) of per- and polyfluoroalkyl substances, PFAS-based films while remaining eco-compliant and durable under abrasion and humidity cycling. Scratch resistance is another key requirement: hard coats with pencil hardness $\geq$ 3H (and up to 9H for premium glass-like finishes) are achieved by cross-linked organosilicate networks or silica nanoparticle-reinforced polymers cured under UV or plasma. In tactile devices, coatings must also preserve low-surface friction (static friction coefficient < 0.3) for smooth gesture input. Self-healing chemistries—incorporating dynamic covalent bonds (Diels–Alder, disulfide exchange) or supramolecular hydrogen bonding—are increasingly integrated to restore surface integrity after microscratches. These systems can recover ~ 70–90% of hardness and gloss after mild heat (60–80 °C) or UV exposure, extending device lifetimes. Additionally, for advanced touch and gesture recognition, the surface must maintain dielectric stability (dielectric constant < 5 and low loss tangent) to avoid interference with capacitive sensing and must minimize haze (< 1%) to preserve resolution and brightness [1, 11–19].

Various expectations of surface functional coating include but not limited to easy application and cost-effectiveness, environmental friendliness, tailored surface morphology, and durability [1]. A key advantage of surface engineering via coating lies in its ability to deliver multifunctionality through carefully engineered material stacks. For instance, a single multilayer interference coating might simultaneously provide anti-reflective, UV-blocking, and anti-static properties. Similarly, nanostructured films inspired by biomimetic design (such as moth-eye textures) can offer low reflectance and high transmittance with minimal added thickness, which is vital for flexible and curved substrates. To maintain longer product life without reducing visual clarity, robust protective layers such as polysiloxane coatings that are scratch, and abrasion resistant are applied to electronic displays to shield its surfaces from everyday wear and tear. Anti-static coatings are used to reduce dust attraction and prevent particles from sticking to display surfaces, thereby minimizing the need for frequent cleaning and reducing yield loss during fabrication caused by debris or surface defects. They also help prevent the buildup of static electricity on the screen, protecting the device from potential damage.

In addition, hydrophobic coating is used to ease keeping the screen surface free of contamination from fingerprints, water/moisture droplets which scatter/distort images and dirt from exposure to outside world, and ability to wipe the surface with soft cloth or tissue. This hydrophobic coating is done by applying a low-surface-energy coating as the topcoat of the display surface such that contact angle of water is greater than 110° [2]. A typical approach used for this is silica sol–gel dip coating to provide hydrophobic and anti-glare functions together with scratch resistance for display surfaces. Sol–gel coatings through dipping or spin-casting methods are used for plastic substrates such as polymer-based optics used in binoculars and camera lenses, other materials such as polyethylene terephthalate (PET), Zeonex, acrylic, polyimide (PI), polymethyl methacrylate (PMMA), and polyethylene naphthalate (PEN) [3]. Hydrophobic surface can also be achieved using perfluoropolyether polymers modified with organofunctional silanes [4]. Furthermore, physical vapor deposition approach is known to be used to conveniently make multilayer optical coatings. Overall, ensuring that surfaces of the coated layers are kept dry and clean is very important for premium display output irrespective of vacuum processes or by non-vacuum sol–gel deposition techniques.

The rainbow effect—also known as color fringing—is a visual artifact in electronic displays caused by uneven light transmission across the screen, resulting in unintended color separation. This phenomenon arises from complex interactions between the display's structural design, the incident ambient light, and the viewer's angle of observation. It is especially prevalent in LCD technologies, where red, green, and blue color filter layers can become misaligned with liquid crystal orientations, leading to color bleeding at pixel boundaries. Additionally, the intrinsic limitations in viewing angles of LCDs can result in angular light distortion, producing pronounced rainbow-like patterns in high-brightness or high-contrast scenes. In LED-based displays, non-uniform backlighting or spectral inconsistencies in the light source can exacerbate this effect. To mitigate such optical artifacts, advanced rainbow-reduction films have been developed. These typically involve optical-grade polyester films coated with formulations containing polyurethane resin, isocyanate and oxazoline curing agents, and well-dispersed inorganic nanoparticles. By fine-tuning the refractive index of the coated layer—typically between 1.55 and 1.65—these films minimize interference patterns and color dispersion, significantly improving image fidelity and uniformity in display performance [5].

The surfaces of electronic displays demand sophisticated optical engineering to ensure optimal image clarity, visibility, and user comfort. Anti-reflection (AR) coatings are indispensable for minimizing specular reflectance that causes image blurring and directs unwanted light into the user's eyes. In parallel, anti-glare (AG) coatings serve to diffuse intense ambient light that can obscure screen visibility, thereby enhancing contrast and legibility under various lighting conditions. Often, a hybrid approach combining both AR and AG functionalities is employed to deliver a balanced solution that supports both clarity and glare mitigation. AR coatings are typically constructed from non-absorbing oxide materials tailored for glass substrates and designed to deliver broadband suppression of reflection across the visible spectrum. High-refractive-index materials such as Ta_2O_5, TiO_2, ZrO_2, and Nb_2O_5 are commonly used in conjunction with intermediate-index layers like Al_2O_3 and low-index materials such as SiO_2 or MgF_2. Standard AR stacks generally comprise three to five layers, whereas advanced imaging systems (e.g., smartphone displays and camera optics) may utilize multilayer structures of up to eight layers to achieve sub-percent reflectance for optimal performance [6]. AG coatings are often realized via surface texturing or incorporation of microparticles—such as silica, PMMA, or acrylic resins—that create controlled surface roughness on the scale of tens to hundreds of nanometers. This microroughness induces diffuse reflection, scattering ambient light across multiple angles to suppress glare. However, such scattering can degrade image sharpness and resolution. To mitigate this drawback, advanced interference-based AG coatings have been developed using both dielectric multi-layers and light-absorbing materials to preserve transmittance while maintaining matte-like anti-glare performance. These may slightly reduce total light transmission but offer a more controlled balance between glare suppression and image clarity [7]. Representative material systems integrating AR/AG functionalities include TiN_xW_y/SiO_2, Cr/CrO_x, Fe/FO_x, ITO/SiO_2, and $TiNO_x/SiO_2$ [8, 9]. The integration of these functional coatings not only ensures high visual acuity and optical uniformity but also reduces visual fatigue, making them vital for both consumer-grade displays and high-precision imaging environments [10].

Other essential optical coatings include infrared (IR) blockers and antimicrobial finishes. IR-blocking coatings are designed to shield electronic displays—such

as outdoor automated teller machines (ATMs) and point-of-sale (POS) terminals—
from solar heat and infrared radiation, thereby enhancing thermal stability and display
longevity in high-temperature environments. Meanwhile, antimicrobial optical coatings, often formulated with silver-ion additives, serve as protective surface finishes
to inhibit the growth and transmission of microbes. These coatings are particularly
important for displays used in sensitive environments such as hospitals, pharma-
cies, bioscience facilities, and food packaging industries, where preventing fungal,
bacterial, and viral contamination is critical [10].

3.1 General Deposition Techniques for Functional Optical Coatings

Tailored functional coatings for electronic displays are achieved through a variety
of sophisticated deposition techniques, each selected based on the desired functional attributes—such as optical transmittance, glare suppression, reflection control,
mechanical durability, thermal stability, and surface resistivity. These thin-film coatings play a vital role in defining and regulating how light interacts with the display
surface, ultimately determining visual performance, environmental resilience, and
device longevity. The selection of a suitable deposition method is critical, as it
governs the microstructural characteristics, thickness uniformity, adhesion strength,
and overall functional integrity of the applied coating. Table 3.1 summarizes typical
thickness ranges, deposition rates, substrate scalability, and uniformity tolerances
for the major deposition techniques based on common engineering practices toward
practical manufacturability in large-area display fabrication [11–20].

Sol–gel deposition remains one of the most versatile approaches, particularly for
producing uniform, large-area coatings at relatively low cost. It involves the use of a
stable colloidal suspension of nanoparticles, referred to as a "sol," which undergoes
controlled hydrolysis and condensation reactions to form a gel-like network. This gel
eventually solidifies into a thin, continuous film upon drying and thermal treatment.
Sol–gel methods are compatible with dip coating, spin coating, spray deposition,
electrochemical deposition, and chemical bath deposition (CBD). Thicknesses per
cycle typically range from 50–500 nm, with deposition rates of 10–200 nm/min
depending on solvent evaporation and annealing. These techniques scale well to
meter-wide glass sheets and flexible polymer webs, making them attractive for cost-
sensitive AR/AG layers and self-cleaning coatings [21, 22].

Physical vapor deposition (PVD) is a high-precision vacuum-based technique
that includes evaporation and sputtering. In PVD, the coating material is vaporized
from a solid or liquid source either thermally or by energetic plasma discharge (sputtering), then condensed onto the substrate to form dense, adherent films. PVD variants
include magnetron sputtering, high-power impulse magnetron sputtering (HiPIMS),
ion beam sputtering (IBS), ion-assisted deposition (IAD), electron-beam evaporation, arc evaporation, and pulsed-laser ablation. Typical film thicknesses range from

5 nm to 5 μm with deposition rates of 0.1–1 μm/h for sputtering and faster for evaporation. Industrially, PVD supports Gen 8.5–10.5 glass panels (>3000 × 3000 mm) with excellent ± 1–2 nm uniformity over large areas, making it the de facto standard for ITO electrodes, optical interference stacks, and hard coatings [10, 22].

Chemical vapor deposition (CVD) involves the decomposition or chemical reaction of vapor-phase precursors within a reaction chamber to produce solid films on the substrate surface. This method enables excellent conformal coverage and strong film-substrate adhesion, making it ideal for complex display architectures. Variants include atmospheric-pressure CVD (APCVD), low-pressure CVD (LPCVD), and plasma-enhanced CVD (PECVD). PECVD is particularly important for low-temperature deposition of dielectric layers (e.g., SiO_2, Si_3N_4) on flexible polymer substrates, while APCVD is favored for large rigid glass sheets due to its high throughput. Typical rates are 10–100 nm/min (slower for LPCVD), with ± 2–5 nm thickness uniformity. CVD enables Gen 6–10 glass sheet coverage and conformal coating of 3D architectures, widely used for barrier protection, optical passivation, and encapsulation in OLED/QD displays [8, 10].

Atomic layer deposition (ALD) provides atomic-scale thickness control through sequential self-limiting surface reactions, yielding exceptionally uniform and pinhole-free coatings. ALD films typically range from 1 to 100 nm with deposition rates of 0.1–5 nm/min. Batch and spatial ALD systems now support meter-wide flexible webs, although throughput and cost remain challenges for full display-scale adoption. ALD is increasingly applied for thin encapsulation layers, high-k gate dielectrics, moisture/oxygen barriers, and wide-angle broadband AR coatings that require ± 0.1–0.3 nm per cycle precision [17, 18].

Roll-to-roll and printing techniques (e.g., slot-die coating, spray, inkjet, gravure) enable high-throughput, low-cost deposition on continuous polymer films > 1 m wide at web speeds of 1–10 m/min, achieving thicknesses from tens of nanometers to several microns with ± 5–20 nm uniformity. UV-curable and nanoparticle-loaded inks are increasingly used for oleophobic, AR/AG, and functional barrier coatings on flexible and wearable displays [19, 20].

Overall, the interplay between deposition technique and coating performance remains a cornerstone of innovation in optical coating systems. As display technologies evolve toward thinner, flexible, and more interactive platforms, advanced deposition processes will continue to drive improvements in optical quality, environmental robustness, and long-term reliability. Beyond identifying deposition methods, it is equally important to contextualize their operating scales and industrial feasibility.

3.1.1 Characterization of Functional Coatings

The performance and long-term reliability of functional coatings in electronic displays are critically governed by parameters such as deposition technique, film thickness, microstructural configuration, and compositional uniformity. To ensure that these coatings meet the rigorous optical and mechanical demands of advanced display technologies, a comprehensive suite of characterization methodologies is employed. Surface roughness and thickness measurements are foundational in determining film uniformity and light-scattering behavior, directly impacting image clarity and display brightness. Hardness and microhardness testing evaluate the coating's resistance to mechanical abrasion and deformation, while adhesion assessments quantify interfacial strength and coating durability under physical stress. Microstructural analysis using techniques such as scanning electron microscopy (SEM) and transmission electron microscopy (TEM) reveals the internal morphology and grain structure, providing insight into growth mechanisms and defect densities. Chemical composition and bonding characteristics are identified via spectroscopic tools such as XPS, FTIR, and EDS, crucial for confirming stoichiometry and detecting impurities that may impair optical performance or induce premature failure. Residual stress measurements, often performed using wafer curvature or X-ray diffraction methods, are vital for predicting mechanical stability, especially under cyclic thermal loads. Depth profiling, commonly executed via secondary ion mass spectrometry (SIMS) or glow discharge optical emission spectroscopy (GDOES), uncovers elemental gradients throughout the coating thickness—key for multilayer systems. To simulate real-world operational scenarios, coatings are subjected to dynamic environmental tests involving thermal cycling, humidity exposure, and chemical resistance trials, assessing their robustness against environmental degradation. Service life predictions are supported by accelerated aging protocols, while impact, scratch, and deformation resistance are gauged through mechanical testing. Thermogravimetric analysis (TGA) and dynamic mechanical thermal analysis (DMTA) further offer detailed insights into the thermal decomposition profile, mechanical integrity, and viscoelastic behavior of the coatings across varying temperatures and strain conditions. Together, these multidimensional evaluations provide a holistic understanding of coating performance, guiding iterative improvements in material formulation and processing for high-fidelity, durable electronic display systems [22].

3.2 Requirements for Anti-Reflective Films

Both anti-reflection (AR) and anti-glare (AG) coatings are collectively referred to as anti-reflective films; however, they function via distinct mechanisms to manage light reflection on electronic display surfaces. AR coatings minimize direct reflectivity by manipulating light interference through precisely engineered thin-film layers without introducing surface irregularities, thereby preserving high image clarity. In contrast, AG coatings reduce the intensity of reflected external light through diffuse reflection, achieved by incorporating surface roughness in the tens to hundreds of nanometers, which scatters incident light at various angles. This bumpy surface texture is effective in mitigating glare under ambient lighting conditions, but it often comes at the cost of image sharpness due to scattering-induced optical distortion. Therefore, in applications requiring pristine visual clarity—such as high-resolution displays—AR coatings are preferred despite their relatively higher cost, as they maintain optical precision without relying on surface irregularities [23].

Generally, there are four types of anti-reflective films based on type of display quality requirement. These include (1) AG, anti-glare film which is made of inorganic or organic particles of specific size mixed in hard layer to coat a required roughness, (2) AGS, anti-glare/anti-static film which has conductive nanoparticle added to the AG to feature anti-static function, (3) AG/AR film which combines both AG and AR features by coating high-refractive-index material such as TiO_2 [24] and low refractive index material such as MgF_2, CeF_3 on the AG topcoat, (4) clear AR film with smooth flat surface with no irregularities by using organic and inorganic particles in the hard layer or same structure as the AG/AR in (3) [23].

Typically, to maintain minimum reflectivity in the visible light region, single-layer and two-layer anti-reflective films are adopted to obtain minimum reflectivity of about 550 nm which stands at middle of visible light range, while a three-layer reflective membrane type designed with a multilayer structure can be used to ensure minimal reflectivity in the entire visible light range. Furthermore, based on the structure of the hypo (low) refractive index used in the anti-reflective film layer, classification is divided into plat plane, a moth-eye [25], nanoporous [26, 27], and hollow [28] structures. While planar type has a restriction on the type of material that can be used as low refractive layer index, the remaining structures are adaptable to reduction in effective refractive index of the low refractive layer to give excellent anti-reflective features. For example, the moth-eye structure which is fabricated using nanoprint or holographic lithography possesses a gradient refractive index and the regular spacing of the structure does not scatter light in the visible light region. The nanoporous scaffold is fabricated by reducing the effective refractive index by building air pockets into the planar low refractive layer, and this is prepared using selective etching method or sputtering through phase separation. The hollow type is built via wet coating by incorporating hollow silica inside the low refractive layer to produce resultant low effective refractive index. Furthermore, since anti-reflective film is a topcoat layer, which is the most exposed part of the electronic display, scratch resistance and weather resistance must be well catered. By standard, the coated anti-reflective films should maintain physical properties such as abrasion resistance, pencil hardness, anti-fouling, and adhesion to ensure good application as shown in Table 3.2 [23].

3.3 Anti-Reflective Film Coatings

To add anti-glare or anti-reflection properties to electronic displays, AG and AR are usually deposited as topcoats of the display screens. Fabrication of anti-reflective films is usually done by means of wet coating (solution processible), which is cost-effective and industrial production scalable, and also, this coating process is adaptable to achieve the expected characteristics of film. For any coating approach used, it should be possible to control the irregularities of the surface in the case of AG coatings, and in the case of anti-reflection coatings, the refractive index of each layer should be easily adjustable depending on the functional material used.

Usually, there are three major steps involved in manufacturing anti-reflective films. First, use of sol–gel technology and UV-curable acrylate technology to synthesize inorganic and organic filler materials. Next, milling approach is used for the production of the nano-sized inorganic filling components that control refractive index, conductivity, and scratch resistance together with dispersion technology used to disperse these coating fillers evenly and stably in a non-organic substrate. Finally, a proper coating process coupled with suitable drying/curing approach is applied using the synthesized coating materials, to deposit anti-reflective films [23].

3.3.1   Anti-glare (AG) Coating Materials

There are two major approaches used to cast irregularities on a display surface during coating of anti-glare (AG). These include the use of embossing to control surface irregularities and secondly by incorporating fillers. Among various methods of carrying out embossing process, use of embossing rolls is widely used in electronic display to annul glare disturbance while keeping the inherent features of the binder. However, this approach is not flexible enough to control the degree of anti-glare, and it is expensive to produce embossing rolls.

Therefore, it is difficult to apply this approach to large-scale production. On the other hand, adoption of fillers is done based on the type of binder and filler dispersion used which usually takes place in three ways. The dispersions include polymer organic filler in polymer binder resin, inorganic filler in polymer binder resin, and inorganic filler & organic polymer filler in polymer binder resin. The binders can be silicone-based thermocurable coatings, polymer-based thermocurable coatings, or acrylate-based UV-curable resins. UV-curable resins are used to improve strength and productivity of the AG film. Dispersion methods like ball mill, dyno mill, ultrasonic grinding are used to add fillers such as organic or inorganic fillers to binders. The required characteristic irregularities in AG film surface during coating and curing of the prepared solution are given by difference in volume shrinkage between the binder and the filler. The degree of irregularities can be controlled by the type, size, content, and coating solution preparation method of the filler. Hence, the degree of anti-glare can be controlled. However, as resolution of display increases, the unevenness of AG films conflicts with high resolution in the latest display products which results in production distorted display images. This can be solved by controlling surface illuminance of AG films. Consequently, various research took place to develop products that have the features of preventing glare via internal scattering by controlling the type, size, and refractive index of the added filler with finer surface irregularities during use in high-resolution electronic display, together with release of carbide products which can significantly improve inner scratch latch properties [23].

3.3.2 Anti-reflection (AR) Coating Materials

Anti-reflection films are usually coated on display devices to minimize reflection of wavelengths of light. This helps to eliminate disturbance of reflective rays to the eyes and reduce loss of light due to the reflection. Generally, the coating is done to manage reflection on the interface between air and a surface by introducing a new interface between the surface and the new coating that serves as extra material for light to interact with. Ultimately, this will lessen the total reflection coefficient of the system such that light from the two interfaces (with each interface having smaller difference in refractive indices than the initial interface) undergoes destructive interference.

Anti-reflection (AR) films function as optical barriers that minimize surface reflectivity by incorporating coatings with tailored refractive indices, allowing destructive interference of incident light. These films are critical in high-resolution electronic displays, where they suppress ambient light reflections without compromising image clarity—unlike anti-glare (AG) coatings that achieve light diffusion through surface microtextures. A typical AR film begins with a hard coating layer to enhance scratch resistance and mechanical durability of the substrate, often composed of silica (SiO_2), alumina (Al_2O_3), or hybrid siloxane-based resins. This is followed by a low-refractive index (LR) layer—commonly made from materials like porous SiO_2, magnesium fluoride (MgF_2), or fluorinated polymers—which forms a basic two-layer system capable of reducing reflectance to around 2%. For higher-performance requirements, a three-layer structure is employed, wherein a high-refractive-index layer (such as titanium dioxide, TiO_2; tantalum pentoxide, Ta_2O_5; or niobium pentoxide, Nb_2O_5) is applied over the hard coat, followed by an LR topcoat. This design reduces reflectance to approximately 1% but can introduce undesirable reflected color tints. To mitigate such optical artifacts, precise control over layer thickness (typically in the 50–150 nm range) and refractive index contrast is essential. More advanced four-layer AR structures have since been developed to further suppress reflection and minimize chromatic distortion. These systems often employ nanoparticle-enhanced coatings using blends of high- and low-index oxides suspended in polymeric binders, allowing fine-tuning of the film's optical profile. For example, formulations may incorporate a TiO_2–SiO_2 nanocomposite layer or combine CeF_3 and ZrO_2 in a graded-index architecture. These innovations have enabled reflectivity to be reduced below 0.5% in the visible spectrum, suppressing color fringing and also supporting stringent clarity and luminance standards required by OLED and AMOLED displays in smartphones, tablets, and AR/VR headsets. In most cases, high- and low-refractive index (RI) solutions are synergistically formulated as composite coating systems—using functional fillers dispersed in polymeric binders—to achieve finely tuned optical properties in AR films [23].

For the high-RI solution, UV-curable acrylate-based binders are commonly employed due to their excellent film-forming ability, mechanical robustness, and enhanced productivity during large-scale fabrication. These binders are used to disperse high-refractive-index fillers such as zirconia (ZrO_2), titania (TiO_2), indium tin oxide (ITO), antimony tin oxide (ATO), and aluminum zinc oxide (AZO). In particular, transparent conductive oxides like ITO, ATO, and AZO offer the dual advantage of high optical transmittance in the visible spectrum and anti-static functionality—making them ideal for applications where optical clarity and electrostatic discharge mitigation are critical. To optimize the refractive index and conductivity, precise control over the type, concentration, and dispersion quality of the nanoparticles is essential. Typically, primary particle sizes should range between 10 and 20 nm, while secondary particle agglomerates dispersed in solvent should remain below 100 nm to minimize Mie scattering. Particle sizes exceeding 100 nm increase the risk of visible light scattering, resulting in reduced transparency and film whitening. Therefore, careful optimization of dispersants and solvent composition is crucial to prevent particle aggregation during formulation.

For the low-RI solution, the choice of binder must facilitate control over the final film's refractive index. Heat-curable silicone-based polymers—such as siloxanes or sol–gel derived networks using tetraethoxysilane (TEOS)—are preferred over acrylates due to their inherently lower RI (~1.45 vs. ~ 1.50). These silicone-based binders, when processed via sol–gel chemistry, enable homogeneous incorporation of low-index fillers. Common fillers include fluorides such as magnesium fluoride (MgF_2), sodium fluoride (NaF), and calcium fluoride (CaF_2), which exhibit excellent optical transparency and low intrinsic refractive indices. Additionally, hollow silica nanoparticles are widely used as low-RI fillers because their internal voids further reduce the effective refractive index. As with the high-RI solution, particle size uniformity is paramount—fillers must remain well below 100 nm to suppress light scattering and ensure high optical clarity. Importantly, particle dispersion stability within the low-index matrix plays a critical role in maintaining uniformity and film performance, necessitating careful control over synthesis and processing conditions [23].

3.4 Anti-reflective Film Manufacturing Process

Similar to roll coating process used for the production of polarizers and PDP filters for FDPs, functional surface thin-film coatings such as anti-glare and anti-reflection films are also deposited using same process. Roll coating approaches suitable for optical coatings are generally divided into dry coating and wet coating. Dry roll coating methods include chemical vapor deposition (CVD) and ion beam sputter ring while wet coating methods include microgravure coater, direct roll coater, reverse roll coater, capillary coater, and bar coater. While dry coating is useful for production of high-quality anti-dust products, high cost and slow production speed render the method not very useful for commercially viable FDP. Hence, it is only used for high-quality products. On the contrary, wet coating process which is commercially viable is widely used to deposit optical coating. Compared to other coatings, optical coating film must be made within a coating thickness of less than 0.1% deviation. Hence the need to deposit it with great precision by using the right coating technique. Deviation in thickness outside 0.1% should be avoided because it causes display haze together with surface strains when observed with naked eyes [23].

3.4.1 Microgravure Coater

Microgravure coaters are the most widely known coaters for optical coatings. The thickness of the coating layer is controlled based on ratio of speed of coating roll to the depth of the groove. It is adaptable to minimize contact area between coating solution and the base film due to the small diameter and line speed. Hence, it is possible to make a thin coating. Compared to gravure coater, clean surface smoothness can be achieved because there is no back roll. However, since coating solution is exposed to atmosphere, concentration of the coating solution easily changes due to solvent evaporation.

3.4.2 Capillary Coater

This coating approach uses capillarity to successfully produce ultra-thin films spanning from few to 0.1 microns. Smooth surface coating is achievable since the feed stream is pulsation-free. It also has the advantage of achieving a coating gap of multiple times more than the coating thickness. However, it is not adaptable to use with resin properties, and the coating speed is restricted.

3.4.3 Rod Coater Mayer

When rod coater Mayer is used, the coating solution is usually deposited on the base material with the aid of a rod of small diameter to achieve smooth coating surface with fixed thickness. To obtain a coating of a given thickness, the size of the space between the wire and the substrate is carefully adjusted, and also, the smaller diameter of wire gives smaller coating thickness and vice versa. Furthermore, this coating approach is adaptable to large-area thin-film coatings with a width of 5 m or more.

3.4.4 Roll Coater

This coater consists of four rolls ranging from liquid transfer direction from the liquid supply: F. roll known as fountain roll, A. roll known as applicator roll, M. roll known as metering roll, and B. roll known as backing roll. The bottom space has liquid supply, and it consists of an F. roll, an M. roll, and two partitions. Due to fine velocity difference between liquid supply along direction of travel as roll rotates, magnitude of coating thickness is maintained. Finally, B. roll induces supply of the coating solution to the substrate for thin-film deposition. The coating thickness can be varied by changing the rotational speed ratio of each roll.

3.5 Comparative Benchmark of AR/AG Coating Strategies for Electronic Displays

To provide a clearer comparison of the diverse anti-reflection (AR) and anti-glare (AG) strategies applied in electronic displays, two benchmark tables are presented (Tables 3.3 and 3.4). Table 3.3 systematically contrasts conventional planar low-refractive coatings, bio-inspired moth-eye nanostructures, nanoporous films, hollow silica particle layers, and hybrid AR/AG systems. Key performance indicators include refractive index (RI), reflectance (R) at 550 nm, transmittance (T), haze (H), angular robustness, durability, and environmental stability. This table consolidates representative values, along with their respective advantages and limitations, providing readers with a concise framework for evaluating AR/AG coating strategies across diverse display applications [6, 25, 27, 29–34].

The comparative analysis shows that each approach involves inherent trade-offs. Planar low-refractive (LR) multilayers can achieve excellent clarity with reflectance values below 1% (at 4 layers), though their angular dependence and multilayer thickness control present challenges. Moth-eye nanostructures can suppress hemispherical reflectance to below 0.1% with outstanding broadband angular stability, yet their high fabrication cost and poor abrasion resistance limit scalability.

Nanoporous polymer films deliver broadband transmittance up to 99.7% through tunable porosity, but their fragile pore networks compromise durability. Hollow silica nanoparticles effectively reduce refractive index and haze, though dispersion control remains difficult. Hybrid AR/AG systems integrate multilayer interference with surface textures, balancing clarity, glare suppression, and robustness, albeit with added process complexity.

For display applications, the selection of coating strategy depends strongly on the operational context: high-resolution handheld devices require smooth AR coatings with minimal haze to maintain clarity; large-area public information panels benefit from hybrid or AG approaches to suppress ambient glare; and flexible or wearable displays require nanoparticle-based or nanostructured films with mechanical adaptability. By framing AR/AG performance in terms of both optical metrics and process feasibility, this benchmark provides a practical roadmap for selecting coating strategies aligned with the evolving requirements of next-generation electronic displays.

Table 3.3 Performance benchmarking of AR/AG coating approaches: trade-offs in reflectance, transmittance, haze, and process attributes [6, 25, 27, 29–34]

Approach	Representative performance	Process advantages	Process limitations	Notes
Planar low-refractive index (LR) films	3-layer:1–2% R, 95–98% T, < 0.5% H; optimized 4-layer): < 1% R, 97–99% T, < 0.5% H	Well-established multilayer process; compatible with oxides (SiO_2, MgF_2, Al_2O_3, TiO_2)	Complex thickness control; limited angular performance	Widely used in smartphones/ cameras with good environmental stability
Moth-eye nanostructures	RI 1.0–1.5, < 0.5% R, 99–99.5% T across visible spectrum, < 0.5% H	Bio-inspired broadband performance; excellent angular stability	Fabrication complexity (nanoimprint, holography); durability concerns	Effective for premium and outdoor displays with excellent broadband / omnidirectional viewing angle
Nanoporous polymer films	RI 1.20–1.35, < 1% R, 98.5–99.7% T, (visible range), 0.5–2% H (if pores > ~ 50 nm	Lightweight, tunable porosity via polymer phase separation	Mechanical fragility; solvent sensitivity	Useful for flexible substrates
Hollow silica nanoparticles	RI 1.22–1.35, < 0.8% R, 98.5–99.7% T, 0.5–1.5% H (tunable)	Low scattering if < 100 nm; scalable sol–gel integration	Agglomeration risk; limited hardness	Good candidate for AR on flexible panels
Hybrid AR/AG films	RI 1.3–1.45, < 1% R, 97–99% T with anti-glare properties, 1–5% H (application tuned)	Combines multilayer interference with matte surface (TiO_2/MgF_2 + surface texture)	Trade-off between clarity and glare suppression; reduced transmittance	Applied in tablets, e-readers, and multifunctional screens

To complement the benchmarking of representative AR/AG approaches presented earlier, it is also critical to assess their comparative performance across practical operational axes such as color stability, abrasion resistance, and cost. Table 3.4 provides a concise head-to-head evaluation from operational and economic standpoints such as color stability, abrasion resistance, and cost for multilayer planar stacks, moth-eye nanostructures, nanoporous polymers, hollow silica nanoparticles, and hybrid AR/AG coatings [6, 25, 27, 31, 35–38]. This comparison highlights not only the optical merits of each approach but also the trade-offs in manufacturability and durability, offering a balanced perspective for guiding material and process selection in real-world display applications.

Table 3.4 Operational and economic benchmarking of AR/AG coating approaches: trade-offs in color stability, abrasion resistance, and cost [6, 25, 27, 31, 35–38]

Approach	Color Shift (ΔE/ Angular)	Abrasion Resistance	Cost and Scalability
Planar LR (SiO_2/ MgF_2 Multilayer)	Moderate; angle sensitive due to interference	Good (4H hardness achievable)	Medium – High (vacuum PVD or sputtering; excellent large-area uniformity on Gen 8–10 glass but requires expensive vacuum systems and tight process control)
Moth-Eye Nanostructures	Minimal; excellent angular stability	Poor–Moderate (fragile nanostructures)	High (nanoimprint lithography / holographic mastering; limited high-volume throughput and costly tooling for meter-scale sheets)
Nanoporous Polymer Films	Low, tunable	Poor (low abrasion/ chemical resistance)	Low (solution spin/dip coating; scalable to roll-to-roll and spray processes; very cost-efficient but fragile)
Hollow Silica Nanoparticles	Low	Fair (improved with hybridization)	Low–Medium (sol–gel routes are inexpensive and scalable to large sheets; some cost increase for surface functionalization and multistep drying/ annealing)
Hybrid AR/AG Films	Moderate (texture-dependent)	Good (reinforced with hard coats)	Medium (combines vacuum deposition for AR stack with low-cost lithography or etching for top texture; scalable with moderate capital / operating cost)

References

1. S.K. Ghosh, Functional coatings and microencapsulation: a general perspective, in *Functional Coatings: By Polymer Microencapsulation* (Germany, 2006), pp. 1–28
2. G. McHale, N.J. Shirtcliffe, M.I. Newton, Contact-angle hysteresis on super-hydrophobic surfaces. Langmuir **20**, 10146–10149 (2004)
3. D. Chen, Y. Yan, E. Westenberg, D. Niebauer, N. Sakaitani, S.R. Chaudhuri, Development of anti-reflection (AR) coating on plastic panels fordisplay applications. J. Sol-Gel Sci. Tech. **19**, 77–82 (2000)

4. L. Mascia, T. Tang, Polyperfluoroether-silica hybrids. Polymer **39**, 3045–3057 (1998)
5. C.J. Pyo, I.S. Jeong, K.J. Moon, Y.S. Kim, *KR102030182B1—Optical Polyester Film with Reduced Rainbow Phenomenon*. Google Patents, South Korea (2013). https://patents.google.com/patent/KR102030182B1/en
6. H.A. Macleod, *Thin-Film Optical Filters*, 4th edn. (CRC Press, 2010)
7. N. Al-Dahoudi, H. Bisht, C. Gobbert, T. Krajewski, A.M. Aegerter, Transparent conducting, anti-static and anti-static–anti-glare coatings on plastic substrates. Thin Solid Films **392**, 299–304 (2001)
8. H. Ishikawa, B. Lippey, Two layer broad band AR coating, in Proceedings of Tenth International Conference on Vacuum Web Coating (Bakish Materials Corporation, New Jersey, USA, 1996), pp. 221–233
9. Y. Zheng, K. Kikuchi, M. Yamasaki, K. Sonoi, K. Uehara, Twolayer wideband antirefl ection coatings with an absorbing layer. Appl. Opt. **36**, 6335–6338 (1997)
10. D. Saddington, *Optical Coatings for Imaging and Displays* (IntertechPira, Maine, USA, 2012)
11. D.L. Smith, D.W. Hoffman, *Thin-Film Deposition: Principles and Practice* (1996).
12. C.J. Brinker, G.W. Scherer, *Sol–Gel Science: The Physics and Chemistry of Sol–Gel Processing* (Academic Press, 1990)
13. C. Sanchez, B. Julián, P. Belleville, M. Popall, Applications of hybrid organic–inorganic nanocomposites. J. Mater. Chem. **15**(35–36), 3559–3592 (2005)
14. D.M. Mattox, *Handbook of Physical Vapor Deposition (PVD) Processing* (William Andrew, 2010)
15. H. Windischmann, Intrinsic stress in sputter-deposited thin films. Crit. Rev. Solid State Mater. Sci. **17**(6), 547–596 (1992)
16. H.O. Pierson, *Handbook of Chemical Vapor Deposition: Principles, Technology and Applications*. (William Andrew, 1999)
17. S.M. George, Atomic layer deposition: an overview. Chem. Rev. **110**(1), 111–131 (2010)
18. M. Leskelä, M. Ritala, Atomic layer deposition (ALD): from precursors to thin film structures. Thin Solid Films **409**(1), 138–146 (2002)
19. F.C. Krebs, Fabrication and processing of polymer solar cells: a review of printing and coating techniques. Sol. Energy Mater. Sol. Cells **93**(4), 394–412 (2009)
20. R.R. Søndergaard, M. Hösel, F.C. Krebs, Roll-to-roll fabrication of large area functional organic materials. J. Polym. Sci., Part B: Polym. Phys. **51**(1), 16–34 (2013)
21. A.J. Olasoji, S.H. Im, Perspective Chapter: TiO$_2$ electron transporting layers for perovskite solar cells [Internet], in *Titanium Dioxide—Uses, Applications, and Advances* (London, UK, 2024). https://doi.org/10.5772/intechopen.1007266IntechOpen
22. D. Thapliyal, S. Verma, K. Tewari, K.C. Bhargava, P. Sen, S. Goel, G. Ghoshal, G.D. Verros, R.K. Arya, Introduction to functional coatings, in *Functional Coatings: Innovations and Challenges* (Wiley, New Jersey, USA; 2024), pp. 1–21
23. Y.J. Hong, Organic/inorganic hybrid coatings for flat panel display. Polym. Sci. Technol. **17**(2), 217–225 (2006)
24. A.J. Olasoji, J.H. Heo, S.H. Im, Facile fabrication of crack-free TiO$_2$ inverse opal thin-film and its application as electron transporting scaffold for efficient Sb$_2$S$_3$-sensitized solar cells. J. Colloid Interface Sci. **678**, 842–853 (2025)
25. A. Gombert, W. Glaubitt, K. Rose, J. Dreibholz, B. Bläsi, A. Heinzel, D. Sporn, W. Döll, V. Wittwer, Subwavelength-structured antireflective surfaces on glass. Thin Solid Films **351**(1–2), 73–78 (1999)
26. J.S. Chapin, Planar magnetron. Vac. Technol. **40**, 37 (1974)
27. S. Walheim, E. Schaffer, J. Mlynek, U. Steiner, Nanophase-separated polymer films as high-performance antireflection coatings. Science **283**(5401), 520–522 (1999)
28. H. Matsuda, T. Ito, Y. Ootani, T. Hirai, M. Kumazawa, M. Komatsu, *A New Concept in Low-Refractive Index, Anti-reflection Coatings Using Porous SiO$_2$ Particles and a Hybrid SiO$_2$ Matrix* (IDW'02, Hiroshima, Japan, 2002), pp. 603–606. Idea and Design Works, LLC (IDW 2002), Japan
29. E.D. Palik, ed., *Handbook of Optical Constants of Solids*, vol. 3 (Academic Press, 1998)

30. O. Stenzel, *Optical Coatings: Material Aspects in Theory and Practice* (2014)
31. P.B. Clapham, M.C. Hutley, Reduction of lens reflexion by the 'moth eye' principle. Nature **244**, 281–282 (1973)
32. M. He, P. Wang, P. Xiao, X. Jia, J. Luo, B. Jiang, Hollow silica nanospheres synthesized by one-step etching method to construct optical coatings with controllable ultra-low refractive index. Colloids Surf. A **670**, 131433 (2023)
33. C. Takai-Yamashita, M. Fuji, Hollow silica nanoparticles: a tiny pore with big dreams. Adv. Powder Technol. **31**(2), 804–807 (2020)
34. S. Bruynooghe, M. Schulze, M. Helgert, M. Challier, D. Tonova, M. Sundermann, T. Koch, A. Gatto, E.B. Kley, Broadband and wide-angle hybrid antireflection coatings prepared by combining interference multilayers with subwavelength structures. J. Nanophotonics **10**(3), 033002–033002 (2016)
35. J.A. Dobrowolski, Optical properties of coatings, in *Handbook of Optical Constants of Solids*, ed by E.D. Palik (Academic Press, San Diego, 2003), pp. 215–260.
36. J.Q. Xi, M.F. Schubert, J.K. Kim, E.F. Schubert, M. Chen, S.Y. Lin, W. Liu, J.A. Smart, Optical thin-film materials with low refractive index for broadband elimination of Fresnel reflection. Nat. Photonics **1**(3), 176–179 (2007)
37. C. Tao, H. Yan, X. Yuan, Q. Yin, J. Zhu, W. Ni, L. Yan, L. Zhang, Sol-gel based antireflective coatings with superhydrophobicity and exceptionally low refractive indices built from trimethylsilanized hollow silica nanoparticles. Colloids Surf. A **509**, 307–313 (2016)
38. Z. Li, C. Song, Q. Li, X. Xiang, H. Yang, X. Wang, J. Gao, Hybrid nanostructured antireflection coating by self-assembled nanosphere lithography. Coatings **9**(7), 453 (2019)

Chapter 4
Practical Applications of Anti-reflective Films in Electronic Displays: Case Studies

This chapter presents five case studies that demonstrate distinct strategies for realizing anti-reflective—anti-reflection (AR) and anti-glare (AG)—functionalities in electronic displays. Each highlights a unique pathway—particle engineering, bio-inspiration, polymer porosity, nanostructured hollow particles, and hybrid designs—offering different balances of optical performance, durability, and manufacturability.

The study by Han et al. [1] shows how polystyrene microsphere-based anti-glare films can be engineered with optimized particle sizes (~3 µm) to achieve uniform scattering surfaces. This approach represents a direct anti-glare film strategy, which eliminates rainbow artifacts, making it attractive for high-resolution displays. Its strengths lie in scalability and robust surface durability, though maintaining transparency at ultra-high pixel densities remains a key limitation.

In contrast, Gombert et al. [2] developed bio-inspired coatings using porous sol–gel processes to replicate subwavelength features of insect eyes which exemplifies moth-eye nanostructure approach. These coatings achieve near-zero reflectance (~0.1%) with exceptional angular stability, representing a benchmark in broadband optical performance. However, the reliance on complex nanofabrication—such as holography or imprint lithography—creates hurdles for mass-scale deployment.

Walheim et al. [3] introduced nanoporous polymer films derived from block copolymer phase separation and selective dissolution which represents nanoporous polymer coating approach. By forming controlled porosity, these coatings reduce effective refractive index and achieve transmission efficiencies exceeding 99%. While they excel optically, their structural fragility and vulnerability to environmental stresses (humidity, abrasion) challenge their long-term reliability in consumer devices.

A. J. Olasoji and S. H. Im, *Functional Coating Layers for Surface Engineering of Electronic Displays*, SpringerBriefs in Applied Sciences and Technology, https://doi.org/10.1007/978-3-032-12843-0_4

Guo et al. [4] reported use of hollow silica nanoparticle coatings, in which closed-pore architectures deliver ultra-low-refractive indices (~1.22–1.30) while improving mechanical durability. This approach represents hollow silica nanoparticle strategy, well-suited for flexible display substrates due to their lightweight, adaptable, however, requires careful sol–gel particle control to prevent agglomeration and scattering.

Finally, Li et al. [5] developed hybrid AR/AG films that integrate metal oxide multilayer interference stacks (TiO_2/SiO_2) with surface nanostructuring. This work typifies the hybrid AR/AG coating approach for glare suppression while preserving high transmittance. Their versatility makes them suitable for tablets, e-readers, and multifunctional screens, though design optimization must carefully balance image clarity with anti-glare efficiency.

Collectively, these case studies illustrate that there is no single universal solution: microspheres prioritize manufacturability and mechanical strength, moth-eye structures set the optical performance benchmark, nanoporous polymers push transmission limits, hollow silica improves the durability–clarity balance, and hybrid AR/AG systems enable multifunctionality. Together, they reveal the breadth of material and structural strategies available to minimize reflection and enhance visual fidelity in next-generation electronic displays, while highlighting the trade-offs in scalability, cost, and robustness.

While Tables 3.1, 3.2 consolidate benchmarking data across major AR/AG strategies, their practical relevance is best illustrated through representative case studies. This chapter presents five examples—microsphere-based AG films, moth-eye nanostructures, nanoporous polymer AR films, hollow silica nanoparticle coatings, and hybrid AR/AG systems—that correspond directly to the approaches benchmarked in Chap. 3. Together, these case studies highlight how laboratory innovations are translated into functional display solutions, underscoring the trade-offs between optical performance, process scalability, and durability in real-world contexts.

Many anti-reflective film strategies are inspired by nature and biological species, such as the compound eyes of moths, gyroid structures in butterfly wings, and other naturally evolved surfaces. Over the years, these biological models have been extensively studied and digitally reconstructed using advanced computational packages, allowing researchers to transform them into engineered optical solutions. As a result, nature-inspired designs have evolved into novel, high-performance anti-reflective approaches for modern display technologies [6].

4.1 High-Resolution Polystyrene Microsphere-Based Anti-glare Film

Han et al. [1] from Im research group, synthesized uniform polystyrene (PS) microspheres in high yield using a continuous long tubular reactor (CLTR) via dispersion polymerization. These PS microspheres were successfully applied in high-resolution display technologies as anti-glare (AG) treatment agents, achieving a total haze value of 16.88, effectively eliminating the undesirable rainbow color effect typically caused by color pixels. Notably, the CLTR system enabled the production of uniform large-sized microspheres (~3 μm) with a high styrene monomer conversion rate of ~ 92%, outperforming conventional batch reactor systems, which produced smaller 1.7 μm particles with a significantly lower yield of ~ 68%.

Smaller PS microspheres (< 3 μm) tend to induce greater surface roughness, resulting in diminished image quality in high-resolution displays (Fig. 4.1a). The superior performance of the CLTR system is attributed to its rapid attainment of reaction temperature, which facilitates the formation of a higher number of nuclei during the early stages of polymerization. Consequently, more monomer molecules were consumed throughout the reaction, leading to the growth of larger PS particles compared to those produced in batch reactors. The synthesized 3 μm PS microspheres (Fig. 4.1b) met the industry standard particle size requirement (3–5 μm) for AG treatments, serving effectively as light-scattering fillers.

The AG film was fabricated by UV-curing a substrate coated with the PS microspheres and UV-curable resins. Upon testing, the resulting film demonstrated excellent mechanical and optical properties, including a pencil hardness of 4H under a 200 g load, and internal, external, and total haze values of 12.50, 4.38, and 16.88,

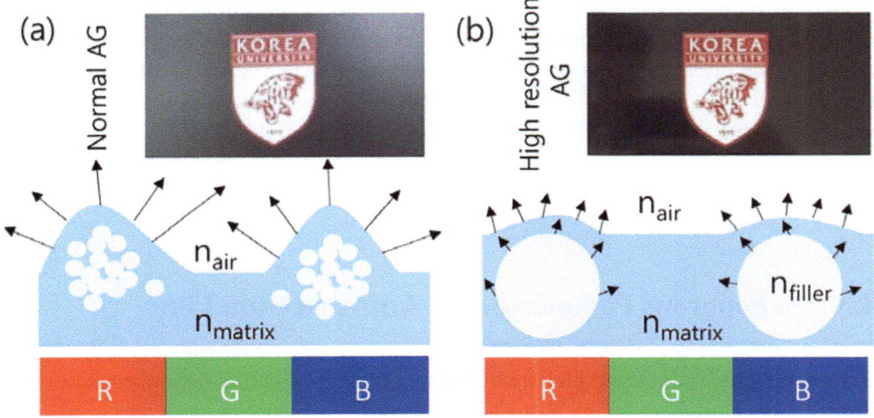

Fig. 4.1 Schematic film structures and typical appearance of the images obtained by **a** normal AG and **b** high-resolution AG film [1]. Reproduced from Ref. [1] with permission from Springer Nature licensed under a creative commons attribution

respectively. These results align well with the standard total haze range of 10–25 required for anti-glare coatings in high-resolution display applications [1].

4.2 Moth-Eye-Inspired-Anti-reflection Film

Gombert et al. [2] developed a low-cost, broadband anti-reflection coating (ARC) using porous sol–gel coatings to fabricate periodic subwavelength surface (SWS) relief structures, drawing inspiration from the corneal surfaces of night-flying moths. These insects naturally evolved lusterless eye surfaces that reduce reflected light, allowing them to avoid detection by predators. Mimicking this bio-inspired mechanism, the researchers engineered nanostructured coatings suitable for application across various display technologies.

The fabrication process began with sol–gel deposition on glass substrates via dip-coating, forming a porous base layer. Subsequently, large-area periodic SWS master structures were generated using holographic exposure of a photoresist. These master patterns were then transferred into nickel molds through electroforming, followed by embossing to replicate the patterns onto organically modified sol–gel materials.

As shown in Fig. 4.2a, scanning electron microscopy (SEM) revealed that the fabricated nanostructures had feature sizes smaller than the wavelength of visible light. Unlike macroscopic structures—which reflect and scatter light due to geometric discontinuities (Fig. 4.2b)—these subwavelength nanostructures interact with light based on Rayleigh scattering principles (Fig. 4.2c). The reduced dimensions allow the nanostructures to bend incident light and guide it into the surface structure, leading to multiple internal reflections and gradual refractive index transition (Fig. 4.2d). This effect significantly reduces surface reflectance.

When the nanostructures are carefully tuned in spacing and depth, they act as effective medium layers that minimize the abrupt change in refractive index at the air–substrate interface. Theoretically, such structures can reduce visible light reflection to as low as 0.1%. Utilizing this method, Gombert et al. reported hemispherical reflectance values of 0.1% for non-absorbing planar substrates—an exceptional performance metric for anti-reflection applications in optical and display systems [6].

4.3 Nanoporous Polymer-Based Anti-reflection Film

Walheim et al. [3] demonstrated a novel method for fabricating nanoporous polymer films as low-refractive-index anti-reflection coatings, which significantly enhanced optical transmittance, making them highly suitable for electronic display applications. The approach relied on the phase separation of macromolecular liquids, specifically a di-block copolymer system.

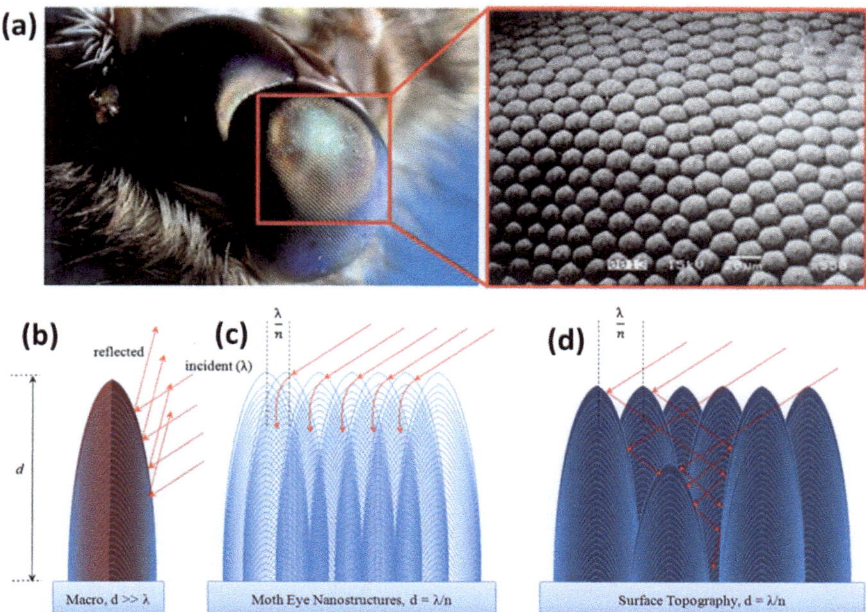

Fig. 4.2 **a** Eye of the moth (left) and SEM image (right) showing the nanostructural sub-wavelength structures on the outer surface of the corneal lens, **b** light being reflected away from a macrostructure, **c** light interacting with the whole rough surface due to comparable dimensions of the microscopic SWS of moth eye, **d** light undergoing multiple internal reflections through moth-eye-inspired and engineered nanostructure pattern [6]. Reproduced from Ref. [6] with permission from the Royal Society of Chemistry licensed under a creative commons attribution

In their process, a low molecular weight polystyrene–polymethyl methacrylate (PS–PMMA) di-block copolymer was first dissolved in tetrahydrofuran (THF) and spin-coated onto both sides of a glass substrate that had been pre-coated with magnesium fluoride (MgF_2), a commonly used high-performance anti-reflective layer. Upon deposition, the resulting film was initially opaque due to the phase-separated morphology of the polymer blend.

To create the nanoporous structure, the coated substrate was immersed in cyclohexane, which selectively dissolves the polystyrene (PS) component, leaving behind a porous matrix of PMMA. This treatment resulted in a transparent nanoporous film with a significantly reduced effective refractive index—ideal for gradient-index anti-reflection behavior.

The final nanoporous polymer coating functioned as a broadband anti-reflection layer with exceptional performance. Notably, the average optical transmittance reached 99.7% across the entire visible spectrum, underscoring the coating's potential for use in high-clarity optical components and advanced electronic display systems [3].

4.4 Hollow Silica Nanoparticle-Based Anti-reflection Film

Guo et al. [4] reported the successful design and fabrication of high-quality hollow closed-pore silica nanoparticle (HSNP) coatings tailored for advanced anti-reflection (AR) performance in optical and display applications. The study addressed one of the persistent challenges of AR coatings—achieving ultra-low-refractive index (RI) values close to air while simultaneously retaining mechanical durability and environmental robustness.

The fabrication approach employed a controlled sol–gel method, in which silica nanoparticles were templated to form hollow, closed-pore structures with precisely engineered shell thicknesses. This hollow geometry effectively reduced the average RI of the coating to values as low as ~ 1.22–1.30, providing a smooth refractive index gradient between air and substrate. As a result, multilayer stacks incorporating HSNP layers achieved reflectance values below 0.5% across the visible spectrum and optical transmittance approaching ~ 99%. Importantly, by carefully controlling particle size (< 100 nm) and shell thickness, the authors minimized light scattering, ensuring transparency and avoiding the whitening effect common in porous systems.

Beyond optics, the study emphasized mechanical and environmental relia-bility. Abrasion testing demonstrated superior hardness compared to conventional nanoporous films, while humidity and temperature cycling confirmed that the closed-pore morphology conferred resistance against water ingress and hydrolytic degrada-tion. These results underscored the importance of combining nanostructural design with mechanical reinforcement to create coatings suitable for consumer displays, outdoor panels, and high-brightness optical devices. By showing that HSNP-based AR coatings could balance ultra-low RI with durability, Guo et al. [4] provided a practical pathway for scalable sol–gel AR films adaptable to flexible and rigid display substrates.

4.5 Hybrid AR/AG Coatings

Li et al. [5] developed a hybrid anti-reflection/anti-glare (AR/AG) coating that syner-gistically integrates multilayer interference design with surface nanostructuring. This approach sought to overcome the limitations of purely planar AR coatings (which provide excellent clarity but poor glare suppression) and traditional AG coatings (which scatter light effectively but compromise image sharpness).

The fabrication process utilized self-assembled nanosphere lithography to produce ordered submicron-scale surface features, combined with multilayer oxide interference stacks (alternating TiO_2 and SiO_2). The multilayer system ensured broadband suppression of Fresnel reflection (< 1% reflectance in the visible range), while the textured topography induced diffuse scattering to mitigate glare under strong ambient illumination. As a result, the hybrid coatings achieved 97–99% optical transmittance, glare suppression tunable between 1 and 5% haze (depending on surface roughness design), and angular robustness superior to planar multilayer AR stacks.

Performance validation demonstrated not only improved viewing comfort under bright light but also mechanical resilience. The coatings, reinforced with hard layers, exhibited pencil hardness ratings up to 4H, making them resistant to scratches and suitable for portable devices subjected to frequent handling. Moreover, the hybrid design minimized color shift ($\Delta E^* < 2$) across wide-viewing angles, which is critical for high-resolution displays such as tablets, e-readers, and multifunctional consumer electronics.

The authors highlighted that the hybrid concept is inherently application-tunable: surface texture parameters can be adjusted to optimize the balance between clarity and glare suppression, while the multilayer design can be tailored for specific spectral ranges. This versatility positions hybrid AR/AG coatings as a leading candidate for multifunctional display surfaces in both indoor and outdoor environments [5].

Representative approaches—including microsphere-based anti-glare films, moth-eye nanostructures, nanoporous polymer coatings, hollow silica nanoparticle composites, and hybrid AR/AG stacks—are compared across key operational parameters in Table 4.1. To complement the theoretical benchmarking of AR/AG strategies presented earlier, this section consolidates real-world case studies that highlight how different coating concepts are practically implemented in display systems [1–5]. The intent is not only to showcase their individual performance metrics but also to frame the trade-offs in terms of optical clarity, environmental durability, scalability, and fabrication feasibility. By situating laboratory innovations within industrially relevant contexts, the table provides a balanced perspective that bridges materials design with practical display engineering. This benchmarking underscores how diverse material and structural strategies converge on the same fundamental goal: reducing reflection and glare while preserving image quality and reliability in advanced electronic display technologies.

Table 4.1 Comparative benchmarking of AR/AG case study approaches in electronic displays [1–5]

Case study (Approach)	Optical performance	Durability	Manufacturability/ scalability	Notes
Han et al.—Polystyrene microsphere AG films	Haze ~ 16–18%; effective glare suppression; eliminates rainbow effect	Good (4H hardness)	High; scalable via UV-curable microsphere embedding	Well-suited for high-resolution displays, but transparency trade-off at ultra-dense pixels
Gombert et al.—Moth-eye	Hemispherical reflectance ~ 0.1%; broadband, angle-stable	Moderate; nanostructures fragile under abrasion	Low–Medium; relies on nanoimprint or holography	Best optical clarity; fabrication cost /complexity high
Walheim et al.—Nanoporous polymer films	> 99% T; RI tunable 1.20–1.35	Low; solvent sensitivity and fragility	Medium; solution-based, scalable	Exceptional clarity, but environmental/ mechanical durability limited
Guo et al.—Hollow silica nanoparticle films	R < 0.5–0.8%; T ~ 99%; controlled haze 0.5–1.5%	Fair; can be improved via hybridization	Medium; sol–gel scalable if agglomeration controlled	Balances clarity and robustness; good candidate for flexible panels
Li et al.—Hybrid AR/AG multilayers	R < 1%; T ~ 97–99%; haze 1–5% (tunable)	Good; reinforced with hard coatings	Medium; PVD + surface texturing	Multipurpose (AR + AG); clarity vs glare suppression trade-off

References

1. J. Han, M.S. You, B.J. Park, S.H. Im, High yield synthesis of polystyrene microspheres by continuous long tubular reactor and their application to antiglare film for high resolution displays. Macromol. Res. **26**, 1095–1098 (2018)
2. A. Gombert, W. Glaubitt, K. Rose, J. Dreibholz, B. Bläsi, A. Heinzel, D. Sporn, W. Döll, V. Wittwer, Subwavelength-structured antireflective surfaces on glass. Thin Solid Films **351**(1–2), 73–78 (1999)
3. S. Walheim, E. Schaffer, J. Mlynek, U. Steiner, Nanophase-separated polymer films as high-performance antireflection coatings. Science **283**(5401), 520–522 (1999)
4. Z. Guo, H. Zhao, W. Zhao, T. Wang, D. Kong, T. Chen, X. Zhang, High-quality hollow closed-pore silica antireflection coatings based on styrene–acrylate emulsion@ organic–inorganic silica precursor. ACS Appl. Mater. Interfaces **8**(18), 11796–11805 (2016)
5. Z. Li, C. Song, Q. Li, X. Xiang, H. Yang, X. Wang, J. Gao, Hybrid nanostructured antireflection coating by self-assembled nanosphere lithography. Coatings **9**(7), 453 (2019)

6. H.K. Raut, V.A. Ganesh, A.S. Nair, S. Ramakrishna, Anti-reflective coatings: a critical, in-depth review. Energy Environ. Sci. **4**(10), 3779–3804 (2011)

Chapter 5
Material Properties of Optical Coatings and Impacts on Optical Performance in Electronic Display Systems

This chapter highlights how material properties form the basis of optical coating effectiveness in display technologies. It begins with the discussion of transparency, clarity, and refractive index control, which are fundamental to managing light transmission, suppressing Fresnel reflection, and maintaining high image quality. Advanced multilayer and gradient-index architectures are examined for their ability to harness interference effects and optimize optical throughput, while attention is given to polarization control and anisotropy management to preserve color accuracy and viewing-angle stability.

The chapter then shifts focus to mechanical and environmental considerations. Hardness, adhesion, and stress resistance are emphasized as critical for durability, especially in flexible or foldable displays where coatings must endure repeated deformation without fracture or delamination. Environmental stability is addressed with particular emphasis on resistance to UV radiation, moisture ingress, and temperature cycling. The importance of uniform thickness and precise deposition techniques is discussed as a prerequisite for reproducibility and reliability. Emerging approaches such as self-healing and smart coatings are introduced as promising strategies to extend display lifetime by enabling coatings to autonomously repair surface damage or adapt to environmental fluctuations.

Finally, the chapter examines characterization and predictive tools alongside real-world applications. Techniques such as ellipsometry, AFM, SEM, and spectrophotometry are described as essential for quantifying optical and mechanical performance, while simulation platforms like RCWA and FDTD allow optimization before fabrication. Practical impacts on display technologies are reinforced through examples, accompanied by analysis of common coating failures and preventive strategies. The chapter concludes with attention to coating–substrate interactions and adhesion chemistry, before looking ahead to innovations such as metasurfaces and bio-inspired designs, which are

A. J. Olasoji and S. H. Im, *Functional Coating Layers for Surface Engineering of Electronic Displays*, SpringerBriefs in Applied Sciences and Technology, https://doi.org/10.1007/978-3-032-12843-0_5

expected to further expand the role of functional coatings in future display systems.

Optical coating plays a pivotal role in enhancing the performance and durability of electronic display technologies. As displays evolve in complexity—ranging from rigid, flat panels to flexible, foldable devices—coatings have become critical in managing light transmission, reducing reflections, providing mechanical protection, and maintaining color fidelity under diverse environmental conditions. The intrinsic properties of the materials used in these coatings govern not only their optical behavior but also their mechanical robustness and environmental stability. Understanding these material properties and their interaction with light and substrates forms the foundation for designing coatings that can meet the stringent requirements of modern displays. This chapter delves into the fundamental material characteristics that determine the optical effectiveness of coatings and explicates their consequential impact on display technology [1].

5.1   Optical Transparency and Clarity

The foremost functional requirement of any optical coating for displays is to maintain high transparency, allowing maximum passage of visible light while minimizing absorption and scattering. Achieving high optical clarity necessitates materials with minimal intrinsic absorption across the visible spectral range (~ 400–700 nm) [2, 3]. Transparency is often quantified by the total transmittance and haze measurements, with high-performance coatings typically exhibiting over 90% transmittance and haze values below 1%. Materials such as silicon dioxide (SiO_2), magnesium fluoride (MgF_2), and certain hybrid organic–inorganic compounds are extensively employed due to their wide optical windows and compatibility with diverse substrates.

Apart from intrinsic material transparency, surface and interface roughness critically affect clarity. Nanometer-scale irregularities on coating surfaces cause diffuse scattering, which increases haze and reduces image sharpness. Uniform film deposition processes that produce smooth, defect-free layers are essential to maintain clarity. Additionally, the coefficient of thermal expansion mismatch between coating and substrate must be minimized to avoid microcracks during temperature fluctuations, which would degrade optical performance.

The interplay between molecular structure and optical properties also guides material choice. For example, materials with low density of electronic states in the visible range reduce photon absorption, thereby increasing transparency. Furthermore, the use of dopants or nanocomposites can tailor optical bandgaps and refractive indices without compromising clarity, enabling multifunctional coatings.

5.2 Refractive Index and Anti-reflective Coatings

Refractive index (n) is a fundamental optical property dictating how light propagates through a material. Differences in refractive indices at interfaces between air, coating layers, and substrates result in partial reflection governed by Fresnel equations. Optical coatings exploit controlled refractive index layering to suppress these reflections via destructive interference.

Single-layer anti-reflective coatings typically utilize materials with refractive indices close to the geometric mean of the substrate and air ($n \approx \sqrt{(n_{substrate} \times n_{air})}$). Such a coating with an optical thickness of one-quarter the target wavelength ($\lambda/4n$) introduces a phase shift causing reflected waves to destructively interfere, thereby minimizing reflected intensity at that wavelength.

More sophisticated multilayer stacks alternate high- and low-index materials to broaden the wavelength range of minimized reflectance. High-index materials include titanium dioxide (TiO_2, $n \sim 2.4$), while low-index layers often comprise silica (SiO_2, $n \sim 1.45$). The precise thickness and order of layers are optimized using optical simulation tools to achieve broadband anti-reflection under various incident angles.

Besides reflection suppression, refractive index engineering influences color rendition and viewing-angle performance. Index-matched coatings reduce chromatic aberrations and preserve image clarity across the panel. Moreover, minimizing refractive index contrast at interfaces reduces stray reflections that degrade contrast ratios and increase ambient light glare, which is particularly important in outdoor and automotive display applications [1].

5.3 Polarization Control and Optical Anisotropy

The manipulation of light polarization is integral to several display technologies, including liquid crystal displays (LCDs), augmented reality (AR) systems, and stereoscopic 3D displays. Optical coatings tailored for polarization control achieve this by exhibiting anisotropic optical properties—differing refractive indices along orthogonal directions—or by incorporating subwavelength structured elements. Polarizers embedded within coatings selectively transmit light oscillating along one axis while absorbing or reflecting orthogonal components. Wire-grid polarizers consist of periodic metallic nanostructures fabricated with pitches smaller than the wavelength of visible light. They efficiently polarize transmitted light, enhance contrast, and reduce ghost images in 3D or polarized displays. Retardation films, such as quarter-wave and half-wave plates, introduce controlled phase delays between orthogonally polarized components, converting linear polarization states into circular or elliptical states, or compensating for birefringence caused by other optical components or mechanical stresses. These coatings often utilize aligned liquid crystal polymers or stretched birefringent polymers to achieve the desired phase retardance.

Critical to their performance is precise molecular or structural alignment to maintain consistent phase retardation across the entire viewing area and under flexing or bending conditions, especially relevant in flexible and wearable displays. Additionally, the coatings must maintain high transparency and low haze while providing stable optical anisotropy over the device lifespan [4].

5.4 Mechanical Properties and Durability

Beyond optical considerations, the mechanical robustness of optical coatings is paramount for ensuring longevity and consistent performance of electronic displays. Mechanical properties such as hardness, elastic modulus, adhesion strength, and flexibility determine how well a coating withstands environmental stresses including abrasion, impacts, bending, and thermal cycling. Hardness relates to coating's resistance to surface deformation and scratching. Materials like diamond-like carbon (DLC) and certain metal oxides (e.g., Al_2O_3, TiO_2) exhibit high hardness values, enabling protection against sharp objects or particulate abrasion common during everyday device use. However, high hardness alone is insufficient; coatings must also maintain adequate toughness to prevent brittle cracking under impact or flexural stress. Elastic modulus and strain tolerance are crucial for flexible displays, which undergo repeated bending and stretching. Coatings composed of polymeric materials or hybrid organic–inorganic composites offer superior elasticity compared to brittle inorganic films. These materials can dissipate mechanical energy without fracturing, preserving optical integrity. Mechanical mismatch between the coating and substrate can lead to delamination or microcracking. Thus, materials are often engineered or layered to optimize gradient mechanical properties, reducing interfacial stresses. Adhesion strength is a fundamental property influencing coating durability. Effective bonding mechanisms include covalent bonding, hydrogen bonding, and physical interlocking. Surface preparation techniques such as plasma treatment or silane coupling agents enhance adhesion by increasing surface energy and chemical compatibility. Without robust adhesion, mechanical stresses induce peeling or blistering, severely degrading optical performance and exposing the underlying substrate to damage. Environmental durability—resistance to humidity, temperature extremes, and chemical exposure—also intersects with mechanical properties. For example, moisture ingress can plasticize polymer coatings, reducing hardness and modulus, whereas thermal cycling induces expansion/contraction cycles that stress interfaces. Coatings are therefore designed with barrier properties and chemical inertness to maintain mechanical performance over operational lifetimes [5, 6].

5.5 Environmental Stability

Optical coatings deployed on electronic displays are regularly subjected to diverse environmental stressors that can undermine their optical and mechanical integrity. Key environmental factors include ultraviolet (UV) radiation, moisture, temperature fluctuations, chemical exposure, and particulate contamination. UV radiation from sunlight or artificial sources initiates photochemical reactions in many polymeric coatings, resulting in bond scission, crosslinking, or oxidation. These processes cause yellowing, embrittlement, and reduced transparency, severely impacting display readability and aesthetics. To mitigate UV degradation, coatings incorporate UV absorbers (e.g., benzotriazoles, hindered amine light stabilizers) or nanoscale inorganic particles like zinc oxide (ZnO) and titanium dioxide (TiO_2), which scatter and absorb harmful wavelengths without affecting visible light transmission. Moisture ingress presents challenges by swelling or hydrolyzing sensitive coating materials, facilitating microbial growth, or corroding underlying electronic components. Hydrophobic coatings, often fluorinated polymers or silane-treated surfaces, provide water repellency and reduce moisture uptake. Additionally, multilayer barrier stacks employing alternating organic and inorganic films create tortuous diffusion paths that dramatically slow moisture permeation. Thermal stability is critical as displays experience wide temperature ranges during operation and storage. Differential thermal expansion between coating layers and substrates induces mechanical stress that can provoke cracking or delamination. High glass transition temperature (T_g) polymers and thermally stable inorganic materials are preferred. Accelerated aging studies under thermal cycling simulate real-world conditions to screen coating formulations. Chemical resistance is vital to protect against oils, cleaning solvents, skin oils, and industrial pollutants. Chemically inert coatings resist swelling, dissolution, and chemical reactions that alter refractive index or surface morphology. Hard inorganic films or fluoropolymer topcoats are commonly used for this purpose. Finally, particulate contamination and abrasion cause surface roughening, increasing scattering and haze. Self-cleaning or oleophobic coatings based on nanotextured surfaces or fluorinated chemistries reduce adhesion of contaminants and facilitate cleaning, thereby preserving optical clarity [7].

5.6 Thickness Uniformity and Deposition Techniques

The thickness and uniformity of optical coatings critically influence their performance, especially when interference effects are exploited to control reflection, transmission, or polarization. Precise control over film thickness at the nanometer scale is mandatory to achieve the designed spectral responses and optical functions. Non-uniform thickness leads to spatial variations in optical properties across the display, causing uneven color shifts, brightness variations, and angular dependence anomalies. Uniformity is influenced by deposition method, substrate geometry, and process parameters. Common deposition techniques include physical vapor deposition (PVD) methods such as electron beam evaporation and sputtering, chemical vapor deposition (CVD), atomic layer deposition (ALD), and solution-based processes like spin coating and dip coating. PVD methods provide dense, high-purity films with excellent thickness control and reproducibility. ALD enables atomic-scale thickness precision and conformal coatings on complex three-dimensional surfaces, increasingly important for flexible or curved displays. Solution processes offer scalability and low cost but often yield less dense films with increased surface roughness. Advances in ink formulation and deposition control have improved their competitiveness. In situ monitoring techniques like quartz crystal microbalance and optical monitoring systems help ensure target thickness and uniformity during deposition. Post-deposition characterization using ellipsometry, profilometry, and spectrophotometry verifies film properties and guides process optimization. The choice of deposition technique balances the required optical performance, mechanical properties, cost, and compatibility with underlying substrates and device fabrication steps [8, 9].

5.7 Flexibility and Adaptation to Emerging Form Factors

With the proliferation of foldable smartphones, wearable devices, and curved automotive displays, optical coatings must adapt to substrates that experience mechanical deformation. Traditional brittle inorganic coatings are inadequate for such applications, prompting the development of flexible, stretchable, and even self-healing optical coatings. Flexible coatings often comprise polymeric materials with inherently high elongation at break and low Young's modulus, enabling them to bend, twist, and stretch without cracking. Materials like polyurethane, polyimide, and fluorinated polymers are popular choices due to their transparency, chemical resistance, and mechanical compliance. Nanocomposite coatings incorporating flexible polymer matrices embedded with nanoparticles (e.g., silica, titania) can combine optical functionality with mechanical resilience. The nanoscale dispersion avoids scattering while enhancing hardness and barrier properties.

Stress-relief mechanisms at the molecular level, such as reversible bond formation or phase-separated morphologies, improve durability under repeated deformation cycles. Recent research into self-healing polymers leverages dynamic covalent chemistry or supramolecular interactions to restore coating integrity after damage. Coating architecture also evolves; multilayer stacks alternate soft and hard layers to distribute strain, while gradient refractive index profiles minimize optical artifacts induced by bending. Furthermore, manufacturing processes such as roll-to-roll coating, spray coating, and inkjet printing facilitate large-area flexible coating deposition with high throughput and reduced cost. Compatibility with emerging flexible display substrates—ultra-thin glass, metal foils, or plastic films—requires tailoring coating adhesion, thermal expansion, and barrier functions accordingly [10, 11].

5.8 Innovations: Self-healing and Smart Coatings

As electronic displays push the boundaries of performance and longevity, traditional static coatings are increasingly supplanted by dynamic, multifunctional films designed to adapt and recover from damage. Self-healing optical coatings represent a transformative innovation, capable of restoring structural and optical integrity after mechanical abrasion, microcracking, or chemical exposure, thus prolonging device lifetime and reducing maintenance costs.

Self-healing mechanisms in coatings typically rely on reversible chemical bonds, microencapsulated healing agents, or supramolecular interactions. For instance, polymer matrices containing dynamic covalent bonds such as disulfide or Diels–Alder linkages can reform broken bonds upon thermal or light stimulus, enabling autonomous repair. Microcapsules embedded within the coating rupture upon damage, releasing monomers or catalysts that polymerize to fill cracks. These approaches preserve optical clarity by maintaining smooth surfaces and consistent refractive indices post-repair.

Beyond self-healing, smart coatings incorporate stimuli-responsive functionalities, adapting optical properties in real-time to changing environmental conditions or user requirements. Thermochromic or photochromic materials adjust transmission or reflectance in response to temperature or light intensity, offering energy-saving advantages through dynamic glare reduction or brightness control. Electrochromic coatings modulate opacity or color upon applied voltage, enabling privacy features or adaptive displays.

Another promising avenue is the integration of nanomaterials such as graphene, carbon nanotubes, or plasmonic nanoparticles to confer multifunctionality, including electrical conductivity, anti-static behavior, or enhanced UV protection. These nanocomposites can also reinforce mechanical properties while preserving or enhancing optical performance.

Such advanced coatings require precise molecular engineering to balance competing demands—maintaining transparency and low haze while enabling dynamic response and mechanical resilience. Manufacturing methods are evolving to incorporate these functional materials uniformly over large areas, and characterization techniques increasingly focus on in situ monitoring of healing or switching behaviors under operational stresses [12, 13].

5.9 Characterization and Metrology Techniques

Accurate characterization of optical coatings is essential for validating performance, guiding process optimization, and ensuring quality control. Given the nanoscale thicknesses and complex multilayer architectures typical of modern coatings, a suite of complementary metrology tools is employed. Spectroscopic ellipsometry is a premier non-destructive optical technique that measures the change in polarization state of reflected light, providing precise thickness, refractive index, and extinction coefficient data across a wide spectral range. Its sensitivity enables detection of sub-nanometer thickness variations and multilayer stack properties, crucial for interference-based coatings. Atomic force microscopy (AFM) offers nanoscale surface topography mapping, revealing roughness, defects, and morphological features affecting scattering and haze. Combined with scanning electron microscopy (SEM), which provides high-resolution imaging and elemental analysis, these tools elucidate the microstructure and uniformity of coatings. Contact angle goniometry assesses surface wettability, indirectly informing on surface energy, cleanliness, and adhesion potential. This metric is especially relevant for coatings designed to be hydrophobic or oleophobic. Environmental testing chambers simulate temperature, humidity, UV exposure, and mechanical stress cycles to accelerate aging and assess coating durability. Optical spectrophotometers track changes in transmittance, reflectance, and color metrics during these tests. Additionally, advanced in situ characterization methods such as real-time spectroscopic monitoring during deposition or mechanical deformation provide insights into dynamic changes and failure mechanisms. Metrology data feed back into modeling and simulation workflows, enabling iterative refinement of coating design for optimal optical and mechanical performance [2, 14].

5.10   Practical Impact on Display Technologies

The integration of advanced optical coatings has substantially elevated the performance and usability of modern electronic displays across multiple application domains. In smartphones and tablets, anti-reflective and oleophobic coatings enhance outdoor readability by reducing glare and fingerprint smudging, respectively. These coatings also contribute to the vivid color reproduction and contrast that consumers expect. The layered design of coatings in AMOLED displays balances brightness retention with power efficiency by optimizing light outcoupling. Large-area OLED televisions utilize hard, abrasion-resistant coatings that preserve screen integrity during handling and cleaning while maintaining color accuracy and viewing-angle consistency. Automotive displays benefit significantly from coatings that manage polarization and reflection under variable lighting, improving driver safety by enhancing visibility and reducing eye strain. Windshield-integrated head-up displays (HUDs) rely on specialized coatings to control light polarization and transmission, enabling crisp virtual images projected onto the glass. Emerging applications such as augmented and virtual reality (AR/VR) headsets require coatings that selectively manipulate polarization and wavefronts for accurate light steering and image formation, while simultaneously protecting delicate optical components. Flexible and foldable displays represent a frontier where coatings must maintain performance under mechanical deformation. Here, innovations in material chemistry and multilayer architecture are enabling durable, optically superior coatings compatible with emerging device form factors. Collectively, these coatings underpin the seamless user experiences and functional robustness that define contemporary electronic display technologies [15].

5.11 Future Directions: Metasurfaces and Bio-inspired Coatings

The trajectory of optical coating development is increasingly driven by advances in nanophotonics and biomimicry, promising revolutionary capabilities beyond classical thin-film designs.

Metasurfaces—ultra-thin layers patterned with subwavelength-scale features—offer unprecedented control over electromagnetic waves, enabling precise manipulation of phase, amplitude, polarization, and direction of light. By engineering the geometry, orientation, and material composition of nanostructures, metasurfaces can realize flat optical components such as lenses, beam deflectors, or polarization converters with minimal thickness and weight. Their incorporation into display coatings could lead to thinner, lighter devices with enhanced optical functions like holography, augmented reality overlays, or dynamic light modulation.

Bio-inspired coatings take cues from natural nanostructures such as the anti-reflective eyes of moths or the iridescent scales of butterflies. These structures achieve broadband, angle-independent anti-reflection, self-cleaning, and structural coloration through hierarchical nano/microarchitectures. Replicating these designs synthetically provides pathways for environmentally resilient, multifunctional coatings that combine optical efficiency with durability and sustainability.

Furthermore, phase-change materials integrated into coatings enable reversible switching between optical states, facilitating adaptive displays that respond to user input or ambient conditions. Sustainability considerations are also gaining prominence, driving research into eco-friendly, solvent-free coatings and closed-loop manufacturing. The challenge lies in developing advanced materials that meet performance criteria while minimizing environmental footprint. Cross-disciplinary collaboration among materials science, optics, chemistry, and engineering will be essential to translate these visionary concepts into practical, manufacturable coatings that redefine the landscape of electronic displays [16, 17].

5.12 Optical Coatings in Real-World Applications

Understanding the theoretical principles of optical coatings is critical, but observing their deployment in real-world display systems provides tangible insights into their practical relevance. Over the past decade, a wide array of optical coatings has been commercialized to meet the demands of consumer electronics, automotive displays, and immersive devices [18].

5.12.1 Smartphone and Tablet Displays

Flagship mobile devices like Apple's iPhone and Samsung's Galaxy series have integrated multilayer optical coatings to deliver high brightness, anti-glare functionality, and smudge resistance. For instance, Apple adopted dual-function coatings combining anti-reflective layers with oleophobic topcoats, reducing ambient reflection while resisting fingerprint buildup from touchscreen interaction. These coatings are engineered with nanometer precision to minimize interference patterns while maximizing transmission of the OLED backlight.

Samsung, on the other hand, employs multistack coatings tailored for AMOLED displays, designed to optimize outcoupling efficiency. These coatings manage waveguide light losses through refractive index tuning, thereby increasing battery efficiency by requiring less energy for the same level of screen brightness.

5.12.2 Augmented Reality and Virtual Reality Headsets

In head-mounted displays like Microsoft's HoloLens or the Meta Quest series, coatings serve a complex array of functions—beam steering, polarization control, and environmental protection. Waveguide optics in AR headsets rely heavily on polarization-selective coatings that separate image light from ambient light, ensuring high visibility in bright conditions. Multilayer thin films in these systems are deposited on curved, transparent substrates, demanding uniformity and adhesion over complex geometries.

Furthermore, anti-glare coatings are applied to exterior visor surfaces to minimize stray light reflections, which can degrade image contrast and interfere with eye tracking systems. These coatings also include anti-scratch functionality, given the proximity of AR/VR devices to the user's face and constant repositioning.

5.12.3 OLED Televisions

Modern OLED TVs by LG, Sony, and others make use of highly engineered hard coat stacks that provide mechanical protection without diminishing image fidelity. These coatings must transmit over 98% of visible light to maintain high luminance and color accuracy. Simultaneously, they must be durable enough to withstand mechanical cleaning and resist microabrasions that could cause scattering and degrade sharpness.

Furthermore, blue-light filtering coatings are increasingly integrated into TV screens to reduce long-term eye strain for users, a concern particularly relevant for displays used over extended periods.

5.12.4 Automotive Displays and HUDs

Vehicles by manufacturers like BMW, Lexus, and Tesla have adopted advanced optical coatings in dashboard displays and head-up displays (HUDs). Coatings for automotive applications must meet a unique blend of performance criteria: optical clarity under sunlight, resistance to wide temperature swings, long-term UV exposure, and mechanical wear. HUDs, which project visual information directly onto windshields, utilize polarization-based coatings to ensure correct image visibility across viewing angles and under varying lighting conditions. These coatings must maintain precise polarization states while being durable enough to survive the vehicle's lifetime. Moreover, anti-fog and hydrophobic topcoats are added to interior display panels to ensure performance in varying humidity conditions.

5.13 Functional Coating Failures in Electronic Displays: Causes and Preventive Measures

Despite technological advancements, optical coatings are still vulnerable to a range of degradation pathways that can undermine both aesthetic and functional performance. Understanding failure mechanisms is essential for predicting product lifespan, improving material design, and developing quality assurance strategies [19].

5.13.1 Delamination and Adhesive Failure

One of the most critical failure modes in multilayer coatings is delamination, where the coating separates from the substrate. This typically arises from weak interfacial bonding due to surface contamination, improper curing, or thermal expansion mismatch between layers. For instance, inorganic oxide films deposited on plastic substrates may experience stress concentration at the interface during heating or mechanical bending, initiating cracks that propagate into full delamination. Mitigating such issues involves surface treatments (e.g., plasma cleaning or UV-ozone exposure) to improve adhesion and designing interlayers that mediate thermal and mechanical property differences. Molecular adhesion promoters like silanes (e.g., APTMS or GPTMS) are used to chemically tether coatings to glass or polymeric substrates.

5.13.2   Photodegradation and UV Yellowing

Many polymeric coatings degrade under prolonged UV exposure due to photochemical reactions. This results in yellowing, microstructural changes, and a reduction in transparency. Degradation is especially problematic in outdoor displays and automotive applications, where long-term exposure to sunlight is inevitable. Stabilizing additives such as UV absorbers, light stabilizers, and radical scavengers can slow this degradation. Alternatively, inorganic top layers such as TiO_2 or ZnO provide UV shielding while maintaining visible light transparency.

5.13.3   Abrasion and Mechanical Wear

Surface abrasion from cleaning, accidental contact, or environmental particles leads to microscratches and haze formation, especially in touchscreen devices and automotive panels. Once surface roughness exceeds a critical threshold, optical scattering increases, impairing image quality. Hard coatings based on silicon nitride, diamond-like carbon, or reinforced polymers improve abrasion resistance. In dynamic environments, self-healing coatings, which can reflow or re-crosslink under mild heat or mechanical stress, can be used to counteract microdamage.

5.13.4   Moisture Ingress and Hydrolysis

Hydrophilic coatings or improperly sealed layers allow water ingress, leading to hydrolysis or delamination over time. This is a particular concern in humid environments or in marine and industrial applications. Barrier coatings employ multilayer oxide and polymer structures, often deposited via atomic layer deposition (ALD), which limit moisture diffusion by creating tortuous paths. Surface functionalization with hydrophobic chemistry strategy (e.g., fluorinated groups) also reduces water penetration. Accelerated life testing—using protocols such as IEC 60068 or MIL-STD-810—exposes coatings to extreme environmental cycling to simulate long-term failure and guide design improvements.

5.14 Coating–Substrate Interaction and Surface Adhesion Chemistry

The interfacial region between a coating and its substrate plays a defining role in the performance and reliability of the overall structure. Adhesion at this interface depends on surface energy compatibility, interatomic bonding mechanisms, and the morphology of the substrate [20].

5.14.1 Surface Energy and Wettability

High-surface-energy substrates tend to promote better adhesion due to increased wettability by coating precursors. For example, untreated polymers like polyethylene terephthalate (PET) have low surface energy, making it difficult for coatings to form continuous films. Surface modification techniques such as plasma activation, corona discharge, and UV-ozone exposure introduce polar functional groups, improving wettability and facilitating chemical bonding. Contact angle measurement is routinely used to assess surface preparation. A decrease in water contact angle after treatment indicates improved hydrophilicity, correlating with enhanced adhesion potential.

5.14.2 Chemical Bonding and Functionalization

The choice of adhesion promoters or primers is often dictated by the substrate material. Glass or silica-based surfaces benefit from silane coupling agents that can covalently bond with hydroxyl groups on the surface and simultaneously anchor organic coating layers. Aminopropyltriethoxysilane (APTES) and glycidyloxypropyltrimethoxysilane (GPTMS) are common choices, forming stable siloxane linkages with the substrate and offering amine or epoxy groups for further reactions. For polymer substrates, adhesion may be enhanced by introducing interlayers that contain compatible chemical groups or that mimic the mechanical behavior of both the substrate and coating, thereby reducing interfacial stress during bending.

5.14.3 Mechanical Interlocking and Morphology

Beyond chemical interactions, physical anchoring contributes to adhesion. Substrates with controlled roughness allow coatings to mechanically interlock with surface features. This approach is particularly effective for thick or viscous coatings that can conform to substrate topography. In flexible displays, where deformation is common, the interface must tolerate mechanical strain without separation. Engineering adhesion through gradient modulus interfaces and compliant adhesion layers mitigates strain concentration at the interface.

5.15 Simulation and Testing Tools for Coating Optimization

The development of high-performance optical coatings relies not only on empirical formulation and experimental validation but also on robust simulation and testing protocols. These computational and experimental approaches enable optimization of coating properties, prediction of long-term behavior, and validation against industry standards, especially for displays that must meet strict optical and mechanical performance metrics [1, 21].

5.15.1 Optical Simulation Tools

Simulation of optical behavior is essential during the design phase of multilayer coatings, especially where interference phenomena, spectral filtering, or angular dependence must be tightly controlled. Tools like TFCalc, OptiLayer, and FilmStar are widely used to model thin-film stacks. These programs simulate reflectance, transmittance, and absorptance spectra across desired wavelengths and incident angles, based on user-defined layer thickness and refractive indices.

Finite-difference time-domain (FDTD) modeling and rigorous coupled-wave analysis (RCWA) are employed when more complex photonic structures are involved—such as nanostructured surfaces, metasurfaces, or polarization-sensitive coatings. These simulations incorporate material dispersion, geometry, and even temperature or voltage dependence to predict real-world performance.

Moreover, COMSOL Multiphysics provides a powerful environment for simulating multi-physics interactions. For instance, it can couple mechanical deformation (in flexible displays) with optical distortion or model thermal effects on the refractive index of coatings under prolonged illumination.

5.15.2 Mechanical and Environmental Stress Modeling

To simulate mechanical reliability, finite element analysis (FEA) is used to model strain distribution, interfacial delamination risks, and fracture propagation in coatings under bending, impact, or thermal cycling. Simulations are especially important in predicting behavior for foldable screens and wearable devices, where coatings must conform to irregular or deforming surfaces. Thermomechanical modeling incorporates coefficients of thermal expansion (CTE), glass transition temperature (T_g), and interfacial adhesion energy to predict delamination, cracking, or warping over device lifetimes. Moisture and gas permeability simulations use Fick's laws of diffusion in multilayer barrier stacks to assess lifetime protection against water vapor, oxygen, or industrial pollutants. These models are calibrated with empirical values from accelerated aging tests.

5.15.3 Experimental Characterization and Testing

Complementing simulations, experimental techniques are critical for verifying the structure and performance of functional coatings. Spectroscopic ellipsometry provides precise measurements of refractive index (n), extinction coefficient (k), and film thickness with sub-nanometer resolution, serving to validate optical models. Profilometry and atomic force microscopy (AFM) assess surface roughness and morphology, offering insights into scattering potential and confirming film uniformity. Spectrophotometry measures transmission, reflectance, and color coordinates (e.g., CIE, Lab) under standardized illumination conditions, allowing direct comparison between experimental spectra and simulated predictions.

5.15.4 Environmental Reliability Testing

To ensure that functional coatings retain their performance under real-world operating conditions, a range of environmental simulation tools and durability testing equipment are employed during the evaluation process. Among these, UV weatherometers are widely used to expose coating samples to controlled levels of ultraviolet (UV) radiation, temperature, and humidity, effectively simulating prolonged sunlight exposure. This allows researchers to assess photodegradation, discoloration, and changes in optical properties over time. Thermal shock testers cycle samples between extremely high and low temperatures—often at rapid rates—to evaluate thermal expansion mismatches and uncover potential issues such as interfacial delamination, microcracking, or mechanical failure. To test mechanical robustness, abrasion and scratch resistance testers—including standardized methods like the Taber abraser test—are utilized to simulate wear caused by styluses, fingers, cloth, or abrasive particles. These tests quantify changes in surface roughness, haze, and coating integrity after repeated cycles. Additionally, for displays such as OLEDs and colored E-paper, the water vapor transmission rate (WVTR) is a critical parameter. It measures the rate of moisture permeation through barrier coatings, helping determine the coating's effectiveness in protecting moisture-sensitive electronic components. Together, these testing methodologies provide a comprehensive understanding of coating durability, informing both material selection and coating architecture for advanced electronic display applications.

5.15.5   Integration of Simulation and Testing

Modern coating development integrates simulation and testing in a feedback loop. Simulations guide the initial design, while experimental testing reveals real-world deviations due to manufacturing imperfections, material aging, or unexpected interfacial effects. These results feed back into models for iterative optimization, enabling coatings tailored for specific display applications—from foldable OLEDs to rugged automotive HUDs.

5.16   Integration of Advanced Materials and Outlook

The field of optical coatings for electronic displays has evolved far beyond its early focus on simple anti-reflective or hard coatings. Today, it lies at the confluence of advanced material science, nanophotonics, surface engineering, and flexible electronics. The demands of modern displays—in terms of brightness, color accuracy, viewing angle, mechanical robustness, and environmental stability—require coatings that are both functionally rich and precisely engineered [22].

5.16.1   Multi-material and Multilayer Integration

Future coatings will increasingly rely on multilayered architectures, where each layer performs a distinct function—optical, mechanical, or chemical. For instance, a single display may contain a hard base layer for abrasion resistance, a refractive index-graded layer for optical coupling, a hydrophobic topcoat for easy cleaning, and a responsive middle layer for smart functions like electrochromic switching. In addition, hybrid organic–inorganic coatings are gaining traction, as they combine the mechanical compliance of polymers with the optical precision and durability of metal oxides. Nanocomposites, incorporating nanoparticles or 2D materials such as graphene and hexagonal boron nitride, can imbue coatings with enhanced conductivity, flexibility, or self-healing characteristics without compromising transparency.

5.16.2   Nanophotonic and Metamaterial Coatings

Advances in metasurfaces and nanostructured coatings promise revolutionary changes in how light is manipulated at the surface of displays. These structures offer opportunities to replace bulky optical elements (e.g., polarizers or lenses) with ultra-thin functional coatings that enable beam steering, holography, or polarization conversion. Moreover, emerging phase-change materials like GeSbTe (GST) allow active modulation of optical states through thermal or electrical stimulus. Such dynamic coatings are central to next-generation smart displays and augmented reality systems where adaptability is key.

5.16.3 Green Chemistry and Sustainable Coating Technologies

Environmental sustainability is becoming a significant driver in optical coating design. Traditional solvent-based coatings can release volatile organic compounds (VOCs) and produce considerable waste during fabrication. Future coating strategies emphasize water-based formulations and UV-curable systems that reduce environmental footprint, solvent-free deposition methods like ALD or vacuum sublimation, closed-loop manufacturing that recycles precursor gases and minimizes waste, biodegradable or recyclable coating materials for circular electronics. These changes are critical for aligning display technologies with global sustainability goals.

5.16.4 Cross-Disciplinary Collaboration and Future Impact

The future of optical coatings lies in cross-disciplinary integration. Materials scientists, optical physicists, chemists, and display engineers must work together to co-develop coatings that meet ever-more-complex device requirements. Innovations in machine learning and computational materials science are accelerating this process by predicting optimal compositions and architectures from vast design spaces.

Ultimately, as electronic displays become thinner, foldable, more immersive, and more energy-efficient, the coatings that protect and empower them must match this sophistication. Optical coatings will not only preserve screen visibility—they will actively shape how we interact with digital information, blurring the line between physical and virtual worlds.

References

1. H.A. Macleod, *Thin-Film Optical Filters*, 4th ed. (CRC Press, 2010)
2. H. Fujiwara, *Spectroscopic Ellipsometry: Principles and Applications* (Wiley, 2007)
3. M. Born, E. Wolf, *Principles of Optics: Electromagnetic Theory of Propagation, Interference and Diffraction of Light* (Elsevier, 2013)
4. A. Yariv, P. Yeh, A. Yariv, *Photonics: Optical Electronics in Modern Communications*, vol. 6 (Oxford University Press, New York, 2007)
5. B. Bhushan, *Principles and Applications of Tribology* (Wiley, 2013)

6. K.L. Mittal (ed.), *Adhesion Measurement of Films and Coatings* (CRC Press, 2014)
7. A. Ghasemi-Kahrizsangi, J. Neshati, H. Shariatpanahi, E. Akbarinezhad, Improving the UV degradation resistance of epoxy coatings using modified carbon black nanoparticles. Prog. Org. Coat. **85**, 199–207 (2015)
8. M. Ohring, *Materials Science of Thin Films: Deposition and Structure* (Academic Press, 2002)
9. D.L. Smith, D.W. Hoffman, Thin-film deposition: principles and practice (1996)
10. D.H. Kim, J.A. Rogers, Stretchable electronics: materials strategies and devices. Adv. Mater. **20**(24), 4887–4892 (2008)
11. T. Sekitani, T. Someya, Stretchable, large-area organic electronics. Adv. Mater. **22**(20), 2228–2246 (2010)
12. S.R. White, N.R. Sottos, P.H. Geubelle, J.S. Moore, M.R. Kessler, S.R. Sriram, E.N. Brown, S. Viswanathan, Autonomic healing of polymer composites. Nature **409**(6822), 794–797 (2001)
13. M.A.C. Stuart, W.T. Huck, J. Genzer, M. Müller, C. Ober, M. Stamm, G.B. Sukhorukov, I. Szleifer, V.V. Tsukruk, M. Urban, F. Winnik, Emerging applications of stimuli-responsive polymer materials. Nat. Mater. **9**(2), 101–113 (2010)
14. J.I. Goldstein, D.E. Newbury, J.R. Michael, N.W. Ritchie, J.H.J. Scott, D.C. Joy, *Scanning Electron Microscopy and X-Ray Microanalysis* (Springer, 2017)
15. M.E. Sousa, G.P. Crawford, Optical and functional coatings for flexible displays, in *Flexible Flat Panel Displays* (2005), pp. 179–193
16. H.T. Chen, A.J. Taylor, N. Yu, A review of metasurfaces: physics and applications. Rep. Prog. Phys. **79**(7), 076401 (2016)
17. P. Lalanne, P. Chavel, Metalenses at visible wavelengths: past, present, perspectives. Laser Photonics Rev. **11**(3), 1600295 (2017)
18. T. Shi, Q. Jia, H. Wang, R. Chen, Preparation of superhydrophobic and oleophobic antireflective coating with high transmittance. Surf. Coat. Technol. **428**, 127863 (2021)
19. H. Cao, R. Zhang, C.S. Sundar, J.P. Yuan, Y. He, T.C. Sandreczki, Y.C. Jean, B. Nielsen, Degradation of polymer coating systems studied by positron annihilation spectroscopy. 1. UV irradiation effect. Macromolecules **31**(19), 6627–6635 (1998)
20. D.S. Liu, C.Y. Wu, Adhesion enhancement of hard coatings deposited on flexible plastic substrates using an interfacial buffer layer. J. Phys. D Appl. Phys. **43**(17), 175301 (2010)
21. Multiphysics, C.O.M.S.O.L., COMSOL® Software (2014)
22. D.D. Paul, Optical Metamaterials: fundamentals and applications (2010)

Chapter 6
Emerging Electronic Display Systems

This chapter explores the shift from rigid flat-panel displays toward versatile, free-form architectures such as foldable, rollable, and stretchable platforms, which serve both as high-quality visual outputs and interactive multitouch interfaces. Functional coatings emerge as indispensable enablers in these systems, safeguarding optical clarity, tactile comfort, and environmental robustness under continuous deformation and frequent user interaction. To achieve this, coatings must integrate hardness, chemical resistance, and transparency with surface functionalities such as low-surface-energy, anti-fingerprint behavior, and lipophobicity.

Foldable and stretchable displays are highlighted as exemplars of this transition. Foldable platforms require coatings that tolerate repeated bending without cracking or loss of adhesion, while stretchable systems demand films capable of sustaining optical and electrical integrity under extreme mechanical strain. Nanoengineered solutions, including nanoparticle-reinforced films and nanostructured interfaces, are shown to deliver tunable refractive indices, adaptive mechanics, and multifunctionality.

Beyond mechanical adaptability, the chapter also examines emerging coating solutions tailored to new display paradigms. For micro-LED displays, specialized topcoats are essential for managing reflections and thermal stability at luminance levels exceeding 5000 cd/m^2. In QD/OLED hybrids, outcoupling layers and protective barriers are introduced to improve photon extraction while preventing quantum-dot degradation. Self-healing and anti-smudge hybrids integrate dynamic polymer matrices with oleophobic additives to prolong device lifetimes in high-use scenarios. Finally, eco-friendly fluorine-free oleophobic coatings, derived from siloxanes, hydrocarbons, or bio-inspired chemistries, represent a sustainable alternative to PFAS-based films, aligning display manufacturing with environmental compliance.

A. J. Olasoji and S. H. Im, *Functional Coating Layers for Surface Engineering of Electronic Displays*, SpringerBriefs in Applied Sciences and Technology, https://doi.org/10.1007/978-3-032-12843-0_6

The chapter closes by addressing scale-up challenges, emphasizing that emerging coating strategies must be compatible with large-area industrial. Collectively, the discussions demonstrate that the success of next-generation displays depends not only on their substrates and electronic architectures but equally on innovative functional coatings that combine optical performance, durability, and compatibility with novel form factors.

The evolution of electronic displays is increasingly moving beyond conventional flat-panel architectures toward more versatile, form-adaptive systems. Today's displays are predominantly multitouch interfaces that offer not only visual output but also interactive input. This shift has driven the emergence of flexible displayssuch as foldable, rollable, and stretchable displays—collectively termed free-form displays—which are capable of conforming to a variety of surfaces and use cases. Such innovations necessitate advanced functional coating solutions capable of ensuring optical performance, mechanical resilience, and long-term reliability under dynamic mechanical deformation and frequent human interaction.

In these emerging platforms, functional coatings serve as more than just passive barriers—they are engineered to provide active protection and seamless interaction. Coatings must resist scratches, abrasions, and chemical attack while maintaining clarity and touch sensitivity. This is particularly critical in multitouch environments where constant swiping and gesture input are standard. Consequently, next-generation coatings are expected to deliver hardness, durability, anti-fingerprint properties, and low surface energy to preserve both tactile and optical quality. With the rise of gesture-based control and three-dimensional touch input, coatings also play a vital role in enabling precise signal recognition by maintaining stable dielectric properties for capacitive sensing and minimizing scattering that interferes with proximity detection and finger-tracking algorithms. As a result, next-generation coatings must offer not only hardness and durability but also lipophobicity, anti-fingerprint properties, and low surface energy to preserve the visual and tactile quality of the display. In many advanced display stacks, functional coatings are integrated either as a topcoat or within the intrinsic layers of the screen to enhance signal-to-noise ratio and ensure high-fidelity touch response under varying ambient and environmental conditions.

To meet these demands, coatings must remain highly conformable, retaining performance on substrates with extreme curvature or under repetitive strain. This places additional emphasis on flexibility, adhesion to low-energy surfaces, and resistance to mechanical fatigue. Nanoengineered coatings, incorporating nanostructures with tunable refractive indices, are particularly promising due to their thinness, high surface area, and multifunctional adaptability for wearable and foldable devices [1]. Their tunable refractive indices, high surface area, and adaptable mechanical profiles allow for thin, multifunctional films that meet the unique demands of foldable and wearable electronics. While the core functional goals—anti-reflection, anti-glare, moisture resistance, UV protection, and surface robustness—remain consistent with those of traditional flat-panel displays, the technological approach is becoming increasingly sophisticated. Nanoparticle-based coatings, bio-inspired textures (e.g., moth-eye structures), and multilayer interference stacks are being tailored specifically for next-generation display platforms. Additionally, scalable deposition methods such as roll-to-roll processing, inkjet printing, and UV-curing are under active development to support the industrial fabrication of coatings over large, flexible substrates.

Alongside free-form displays, new generations of display systems—including micro-LED arrays, QD/OLED hybrids, self-healing and anti-smudge hybrids, and eco-friendly fluorine-free oleophobic coatings—introduce additional challenges and opportunities. Micro-LED displays require specialized topcoats to minimize reflection, manage heat, and improve pixel reliability under high-brightness operation. Quantum-dot OLED (QD/OLED) hybrids increasingly depend on engineered outcoupling layers that enhance photon extraction efficiency and color purity. Parallel efforts are advancing self-healing and anti-smudge hybrid coatings to prolong device lifetimes in portable, touch-driven environments, as well as eco-friendly fluorine-free oleophobic coatings that replace conventional PFAS (per- and polyfluoroalkyl substances) while retaining strong oil and water repellency. Collectively, these developments highlight how coatings are no longer peripheral finishes (topcoats) but fundamental enablers of next-generation displays, bridging optics, mechanics, and sustainability in a single functional interface.

6.1 Emerging Display Architectures

The following subsections focus on two key free-form display types—foldable and stretchable systems—to illustrate how the advanced coating requirements discussed earlier translate into practical device architectures and real-world engineering solutions.

6.1.1 Foldable Displays

Among the rapidly advancing frontiers of electronic display technologies [1], flexible displays have emerged as a transformative platform, with foldable displays representing one of its most commercially impactful applications [2]. The proliferation of mobile data consumption, driven by high-speed internet infrastructure and continued innovations in mobile device hardware, has fueled a growing demand for larger, more immersive screens within compact and portable devices. Foldable displays provide a viable solution to this design challenge by enabling large-area displays to be embedded within small form factors, effectively combining portability with enhanced visual experiences [3–8]. This advancement is fundamentally enabled by replacing conventional rigid glass substrates with flexible plastic substrates—most notably polyimide (PI)—that can withstand repeated mechanical deformation such as bending, folding, and twisting. Despite their flexibility, these substrates impose limitations due to their lower thermal stability (typically tolerating temperatures up to ~ 500 °C), which restricts the range of compatible coating materials and deposition processes. The first public reports of foldable display concepts trace back to 2005 [9], and by 2019, Samsung Electronics launched the world's first commercial foldable smartphone, pioneering flexible smartphone displays [5]. To ensure optimal optical performance in foldable displays, several functional coatings are integrated into the display architecture. Polarizers are applied to suppress ambient light reflections and enhance visibility, while retarder films are used to adjust the polarization state of emitted light, improving viewing-angle performance. Anti-reflective coatings reduce surface glare and enhance light transmission, and hard coatings improve surface scratch resistance and durability. However, the integration of these coatings must comply with the thermal and mechanical constraints imposed by plastic substrates, requiring the use of low-temperature deposition methods such as solution processing, plasma-enhanced techniques, or vacuum coating under reduced thermal loads. Coating materials are often tailored as thermally stable polymers, nanocomposites, or engineered multilayer films that can maintain transparency, flexibility, and adhesion over prolonged use. Furthermore, foldable displays typically maintain an overall display stack thickness of approximately 0.2 to 1.0 mm, which requires coatings that are optically efficient while also being thin and mechanically compliant to endure repeated folding cycles [10]. Therefore, foldable displays represent a major shift in modern display architecture, and their success hinges significantly on the performance and reliability of functional optical coatings capable of withstanding mechanical stress and maintaining visual fidelity under real-world operating conditions.

6.1.2 Stretchable Displays

Among the various types of mechanically deformable displays—such as foldable, rollable, and stretchable displays—stretchable displays represent the most advanced and versatile form factor [11]. They are regarded as one of the most promising display technologies due to their broad range of potential applications in fields including wearable electronics, fashion, biomedical devices, automotive interiors, gaming, and interactive surfaces [12]. The growing demand for conformable and body-adaptable displays has been further accelerated by the rise of the Internet of Things (IoT) and the increasing interest in continuous physiological monitoring via wearable systems [13]. Stretchable displays enable direct integration of smart displays onto the human body, allowing real-time data collection and visualization through embedded sensors and display modules [14]. These systems must exhibit mechanical softness and stretchability without sacrificing optical or electrical performance, especially in terms of display brightness, resolution, and image quality [15].

Two primary approaches have been adopted in the fabrication of stretchable displays. The first approach focuses on the use of intrinsically stretchable materials for each functional layer of the display—including electrodes, emission layers, transport layers, and encapsulation. Owing to their material compatibility and formability, polymer-based organic light-emitting diodes (OLEDs) have become the dominant technology for such inherently stretchable display systems. In this design, nanoparticle-based optical coatings are also employed due to their ultra-thin-film characteristics, which offer low mechanical stiffness while maintaining light modulation efficiency [16, 17]. The second approach leverages structural engineering of non-stretchable materials to achieve mechanical compliance. Strategies such as island–bridge networks, kirigami, and origami-inspired geometries, buckled structures, and the use of plasticizers or mechanical enhancers allow for mechanical deformation without catastrophic material failure. While these structural designs provide greater material selection flexibility, they face limitations in achieving uniformly distributed strain, necessitating the development of strain-tolerant materials even at moderate deformation levels [15].

Significant milestones in this field include the first demonstrations of stretchable inorganic LED (ILED) arrays by J. A. Rogers' group in 2009 [18] and stretchable OLED arrays by Takao Someya's group the same year [19]. These were followed by various device architectures, including fiber-shaped EL devices, wrinkled EL films, and intrinsically stretchable electroluminescent (EL) prototypes [20–23]. Industry leaders such as LG Display [24] and Samsung Display [25] have also actively developed prototype systems. However, as of now, no commercial stretchable display products have been launched. This is primarily due to unresolved challenges, including suboptimal device efficiency, underdeveloped large-area fabrication processes, and complexity in handling diverse device architectures and materials within a single stretchable platform [12].

6.2 Emerging Coating Solutions

While Sect. 6.1 focused on how free-form display architectures (foldable, rollable, and stretchable systems) redefine mechanical and optical design requirements, an equally important frontier lies in new classes of functional coatings that directly address the unique performance limits of next-generation display stacks. These solutions are no longer confined to only suppressing glare or improving scratch resistance; they are now also engineered to tackle challenges such as extreme luminance, photon extraction losses, surface contamination, and environmental sustainability. Recent innovations include specialized topcoats for micro-LEDs to control reflection and thermal degradation under high brightness, quantum-dot OLED (QD/OLED) outcoupling layers to boost light extraction and color accuracy, self-healing and anti-smudge hybrids that extend service life in touch-intensive devices, and eco-friendly fluorine-free oleophobic coatings that meet tightening environmental regulations while retaining oil and water repellency. Together, these emerging coatings illustrate a shift from passive protection toward active optical and functional engineering, enabling the performance, longevity, and manufacturability of the most advanced display platforms.

6.2.1 Specialized Micro-LED Display Topcoats

Micro-LED technology represents a paradigm shift in display performance, offering unprecedented brightness, long operational lifetimes, and superior energy efficiency compared to OLED and LCD counterparts. However, the extreme luminance levels (often exceeding 5000–10,000 cd/m^2 for outdoor or AR/VR use) introduce new optical and thermal management challenges that must be addressed through advanced coating strategies.

Specialized topcoats are essential to suppress surface reflections, which can otherwise cause viewer discomfort or limit contrast in high-luminance environments. Low-refractive-index AR coatings, often realized via MgF_2 or nanoporous SiO_2, are being tailored with nanostructured texturing to ensure minimal reflection even at oblique viewing angles. Moreover, because micro-LED pixels are fabricated at micron scales, any coating non-uniformity can manifest as color shift or pixel crosstalk. Thus, deposition must achieve sub-nanometer thickness control and excellent lateral uniformity.

Thermal stability is equally critical: coatings must resist yellowing or structural degradation under sustained high-power operation. Hybrid organic–inorganic coatings, particularly those incorporating ceramic nanoparticles into polymer matrices, are being developed to combine optical clarity with thermal robustness. Additionally, hydrophobic and oleophobic top layers are applied to maintain surface cleanliness, which is critical for consumer devices like AR headsets and large public signage that employ micro-LEDs [26, 27].

6.2.2 Quantum-Dot/OLED Outcoupling Layers

Quantum-dot/OLED, QD/OLED displays combine the emissive power of blue OLED subpixels with quantum-dot conversion layers to generate red and green output, thereby improving efficiency and color accuracy. A persistent challenge in these architectures is the limited light outcoupling efficiency—only 20–30% of generated photons typically escape due to waveguiding and internal reflections within the display stack.

Engineered outcoupling coatings are being introduced to mitigate these losses. These include nanopatterned scattering films that redirect trapped waveguided light into the viewing cone, graded-index coatings that minimize refractive index mismatches at interfaces, and quantum-dot encapsulating barrier coatings that both enhance extraction and protect QDs from oxygen and moisture degradation. Materials such as silica aerogels, nanoparticle-loaded resins, and moth-eye-inspired nanostructures are actively being evaluated.

Coating durability is especially important, as QD layers are chemically sensitive and prone to photobleaching. Thus, barrier coatings must combine UV-blocking, low moisture permeability, and mechanical hardness with minimal absorption in the visible spectrum. Advances in atomic layer deposition (ALD) are particularly promising for creating ultra-thin, pinhole-free inorganic protective coatings that simultaneously serve as optical out couplers [28–31].

6.2.3 Self-healing and Anti-smudge Hybrid Coatings

As displays become increasingly tactile—frequently swiped, tapped, and gestured on—the need for coatings that can resist or even self-correct surface damage has become a major research focus. Self-healing coatings integrate reversible chemical bonds (such as disulfide, Diels–Alder, or hydrogen-bonding moieties) into polymer matrices, allowing scratches or abrasions to "heal" under mild heat or light exposure. These coatings extend the functional lifetime of portable devices, reducing the need for frequent replacements.

Anti-smudge functionality is often combined with self-healing via hydrophobic/oleophobic additives (e.g., silane-modified resins or fluorine-free perfluoropolyether analogues) that repel oils and fingerprints. Recent advances focus on hybrid coatings that achieve both resilience and user comfort: multicomponent films where the bulk polymer matrix provides mechanical toughness, while surface-segregating additives migrate to the topmost layer to impart smudge resistance. The interplay between viscoelastic recovery (self-healing) and surface repellency ensures that even heavily used devices can maintain a pristine viewing surface [32–34].

6.2.4 Eco-Friendly Fluorine-Free Oleophobic Coatings

Traditional oleophobic and hydrophobic coatings rely heavily on fluorinated compounds (such as per- and polyfluoroalkyl substances, PFAS), which have come under increasing scrutiny due to environmental persistence and toxicity. The shift toward fluorine-free oleophobic layers represents both a technological and regulatory frontier for display coatings.

Research in this domain explores alternative chemistries such as siloxane-based networks, hydrocarbon waxes, zwitterionic polymers, and bio-inspired hydrophobes derived from fatty acids or plant cuticles. These materials can be engineered to produce low surface energy comparable to fluorinated systems, thereby resisting fingerprints, oils, and stains without environmental drawbacks.

In addition, process scalability is a priority: fluorine-free coatings must be compatible with industrial deposition techniques such as slot-die coating, spray deposition, or UV-curing. Mechanical endurance testing (abrasion, humidity cycling, and chemical exposure) indicates that several hybrid siloxane–hydrocarbon coatings already approach the durability benchmarks of PFAS-based films, making them strong candidates for next-generation eco-compliant consumer electronics [35–39].

6.3   Deposition of Large-Area Optical Coatings for Flexible Displays

As display applications expand into foldable, stretchable, transparent, and large-area formats, functional coatings are increasingly being tailored to meet the emerging requirements of next-generation interfaces, making them central to the future trajectory of display innovation. Gibson et al. [40] successfully carried out low-temperature deposition of multilayer optical coatings suitable for flexible electronic display comprising of many layers of anti-reflection films together with hydrophobic topcoating on a transparent flexible polymer substrate such as polycarbonate and polyethylene terephthalate, PET.

Uniform optical coatings of $0.6\,m \times 1.55\,m$ using a $0.55\,m$ cylindrical drum were deposited at a substrate temperature of $60\,°C$ using reactive closed-field magnetron (RCFM) sputtering process, which is known for the ability to produce dense, uniform, spectrally stable metal-oxide optical films [1].

The AR coating consists of six alternating layers of low- and high-refractive-index materials of SiO_2 and ZrO_2 respectively to enhance transmittance at average reflectance of 0.5% across visible spectrum while the low-surface-energy hydrophobic topcoat was applied using boat sources via thermal evaporation to produce a water-repellant surface with contact angle exceeding $100\,°C$, typical of fluoropolymers such as polytetrafluoroethylene (PTFE) or organosilane-based compounds.

References

1. D. Saddington, Optical Coatings for Imaging and Displays published by IntertechPira 19 Northbrook Dr Portland, Maine 04105 USA (2012)
2. J.H. Back, Y. Kwon, H. Cho, H. Lee, D. Ahn, H.J. Kim, Y. Yu, Y. Kim, W. Lee, M.S. Kwon, Visible-light-curable acrylic resins toward UV-light-blocking adhesives for foldable displays. Adv. Mater. **35**(43), 2204776 (2023)

3. J.C. Lee, S.E. Hudson, E. Tse, Foldable interactive displays, in *Proceedings of the 21st Annual ACM Symposium on User Interface Software and Technology* (Association for Computing Machinery (ACM) publishing, New York, United States, 2008), pp. 287–290

4. C.X. Xu, S. Shu, Q. Yao, G.C. Yuan, H.L. Wu, S. Sun et al., P-117: FLI structure for R1 foldable AMOLED display, in *SID Symposium Digest of Technical Papers*, The Society for Information Display (SID), California, USA, 2020

5. Samsung Newsroom, *Samsung Unfolds the Future with a Whole New Mobile Category: Introducing Galaxy Fold* (2019). http://bit.ly/2SD4yQY. Accessed: February 2019

6. Y. Du, Y. Bai, W. Cao, L. Meng, Y. Bai, Research progress on displays and optical adhesives for flexible 3C products. Eur. Polym. J. 113053 (2024)

7. B. Yuan, J. Li, G. Xu, C.J. Lin, G. Qiao, G. Tao, L. Fu, Q. Shan, 14.2: Invited paper: progress and technical challenges of in-folding display. Sid's Digest Tech. Pap. **52**(S1), 89 (2021)

8. F. Liu, J. Han, J. Qi, Y. Zhang, J. Yu, W. Li, L. Dong, L. Chen, B. Li, Research and application progress of intelligent wearable devices. Chin. J. Anal. Chem. **49**(2), 159–171 (2021)

9. E. Huitema, G. Gelinck, P. Van Lieshout, E. Van Veenendaal, F. Touwslager, Flexible electronic-paper active-matrix displays. J. Soc. Inform. Display **13**(3), 181–185 (2005)

10. E. Huitema, The future of displays is foldable. Inf. Display **28**(2–3), 6–10 (2012)

11. K. Keum, S. Yang, K.S. Kim, S.K. Park, Y.H. Kim, Recent progress of stretchable displays: a comprehensive review of materials, device architectures, and applications. Soft Sci. **4**(4) (2024)

12. H.Y. Zhang, S.X. Jia, D. Yin, J. Feng, Challenges and opportunities of stretchable electroluminescent devices and displays. Laser Photonics Rev. **18**(11), 2400358 (2024)

13. U. Linderhed, I. Petsagkourakis, P.A. Ersman, V. Beni, K. Tybrandt, Fully screen printed stretchable electrochromic displays. Flex. Printed Electron. **6**(4), 045014 (2021)

14. J.H. Koo, D.C. Kim, H.J. Shim, T.H. Kim, D.H. Kim, Flexible and stretchable smart display: materials, fabrication, device design, and system integration. Adv. Func. Mater. **28**(35), 1801834 (2018)

15. Y. Lee, W. Guan, E.Y. Hsieh, S. Nam, Opportunities for nanomaterials in stretchable and free-form displays. Small Sci. **4**(3), 2300143 (2024)

16. C. Wang, K. Xia, H. Wang, X. Liang, Z. Yin, Y. Zhang, Advanced carbon for flexible and wearable electronics. Adv. Mater. **31**(9), 1801072 (2019)

17. A. Qiu, P. Li, Z. Yang, Y. Yao, I. Lee, J. Ma, A path beyond metal and silicon: polymer/nanomaterial composites for stretchable strain sensors. Adv. Func. Mater. **29**(17), 1806306 (2019)

18. S.I. Park, Y. Xiong, R.H. Kim, P. Elvikis, M. Meitl, D.H. Kim, J. Wu, J. Yoon, C.J. Yu, Z. Liu, Y. Huang, Printed assemblies of inorganic light-emitting diodes for deformable and semitransparent displays. Science **325**(5943), 977–981 (2009)

19. T. Sekitani, H. Nakajima, H. Maeda, T. Fukushima, T. Aida, K. Hata, T. Someya, Stretchable active-matrix organic light-emitting diode display using printable elastic conductors. Nat. Mater. **8**(6), 494–499 (2009)

20. D. Hu, X. Xu, J. Miao, O. Gidron, H. Meng, A stretchable alternating current electroluminescent fiber. Materials **11**(2), 184 (2018)

21. Z. Zhang, L. Cui, X. Shi, X. Tian, D. Wang, C. Gu, E. Chen, X. Cheng, Y. Xu, Y. Hu, J. Zhang, Textile display for electronic and brain-interfaced communications. Adv. Mater. **30**(18), 1800323 (2018)

22. Z.Y. Chen, D. Yin, Y.P. Wang, H.Y. Zhang, S.X. Jia, J. Feng, Highly transparent and stretchable organic light-emitting diodes with ultrathin metal films as double electrodes. Appl. Phys. Lett. **122**(5) (2023)

23. M.S. White, M. Kaltenbrunner, E.D. Głowacki, K. Gutnichenko, G. Kettlgruber, I. Graz, S. Aazou, C. Ulbricht, D.A. Egbe, M.C. Miron, Z. Major, Ultrathin, highly flexible and stretchable PLEDs. Nat. Photonics **7**(10), 811–816 (2013)

24. H. Jung, C.I. Park, M.B. Gee, J. Choi, Y.R. Jeong, S.J. Min, J.H. Song, M.S. Lim, M. Kim, T. Kim, S. Ham, High-resolution active-matrix micro-LED stretchable displays. J. Soc. Inform. Display **31**(5), 201–210 (2023)

25. J.H. Hong, J.M. Shin, G.M. Kim, H. Joo, G.S. Park, I.B. Hwang, M.W. Kim, W.S. Park, H.Y. Chu, S. Kim, 9.1-inch stretchable AMOLED display based on LTPS technology. J. Soc. Inf. Display **25**(3), 194–199 (2017)

26. A.R. Anwar, M.T. Sajjad, M.A. Johar, C.A. Hernández-Gutiérrez, M. Usman, S.P. Łepkowski, Recent progress in micro-LED-based display technologies. Laser Photonics Rev. **16**(6), 2100427 (2022)

27. A. Pandey, M. Reddeppa, Z. Mi, Recent progress on micro-LEDs. Light: Adv. Manuf. **4**(4), 519–542 (2024)

28. G. Gomard, J.B. Preinfalk, A. Egel, U. Lemmer, Photon management in solution-processed organic light-emitting diodes: a review of light outcoupling micro-and nanostructures. J. Photonics Energy **6**(3), 030901–030901 (2016)

29. T. Lee, M. Kim, B. Chun, G. Park, S. Yim, S. Yu, J. Kwak, Recent advances in light outcoupling from quantum-dot light-emitting diodes. ACS Photonics **11**(12), 5050–5060 (2024)

30. H. Liang, R. Zhu, Y. Dong, S.T. Wu, J. Li, J. Wang, J. Zhou, Enhancing the outcoupling efficiency of quantum dot LEDs with internal nano-scattering pattern. Opt. Express **23**(10), 12910–12922 (2015)

31. S. Li, L. Lan, M. Li, Z. Gao, X. Yan, D. Fu, X. Sun, Thin-film encapsulation for OLEDs and its advances: toward engineering. Materials **18**(13), 3175 (2025)

32. Z. Wang, L. Scheres, H. Xia, H. Zuilhof, Developments and challenges in self-healing antifouling materials. Adv. Func. Mater. **30**(26), 1908098 (2020)

33. J. Ma, L.E. Porath, M.F. Haque, S. Sett, K.F. Rabbi, S. Nam, N. Miljkovic, C.M. Evans, Ultra-thin self-healing vitrimer coatings for durable hydrophobicity. Nat. Commun. **12**(1), 5210 (2021)

34. J. Zhang, V. Singh, W. Huang, P. Mandal, M.K. Tiwari, Self-healing, robust, liquid-repellent coatings exploiting the donor–acceptor self-assembly. ACS Appl. Mater. Interfaces **15**(6), 8699–8708 (2023)

35. X. Wu, M. Li, X. Chen, P. Xiong, H. Gao, H. Li, S. Zhao, Z. Wang, Z. Wang, Fluorine-free liquid repellent coatings with fully self-healing and robust adhesion abilities. ACS Sustain. Chem. Eng. **12**(38), 14319–14330 (2024)

36. W. Zhang, S. Li, D. Wei, Z. Zheng, Z. Han, Y. Liu, Fluorine-free, robust and self-healing superhydrophobic surfaces with anticorrosion and antibacterial performances. J. Mater. Sci. Technol. **186**, 231–243 (2024)

37. A. van Dam, S.P. Pujari, M.M. Smulders, H. Zuilhof, Fluorine-free hydrophobic polymer brushes for self-healing coatings. ACS Appl. Nano Mater. **7**(16), 19737–19744 (2024)

38. L. Wasser, S. Dalle Vacche, F. Karasu, L. Müller, M. Castellino, A. Vitale, R. Bongiovanni, Y. Leterrier, Bio-inspired fluorine-free self-cleaning polymer coatings. Coatings **8**(12), 436 (2018)

39. H. Wang, H. Zhou, S. Liu, H. Shao, S. Fu, G.C. Rutledge, T. Lin, Durable, self-healing, superhydrophobic fabrics from fluorine-free, waterborne, polydopamine/alkyl silane coatings. RSC Adv. **7**(54), 33986–33993 (2017)

40. D.R. Gibson, I.T. Brinkley, G.H. Hall, E.M. Waddell, A.R. Waugh, J.M. Walls, High performance multilayer optical coatings on flexible sheet, in *Proceedings of The Annual Technical Conference-Society of Vacuum Coaters*, vol. 50. (Society of Vacuum Coaters (SVC), Albuquerque, New Mexico, 2007), p. 319

Chapter 7
Future Trends in Advanced Applications of Commercial Display Technologies

This chapter explores how commercial display technologies are advancing beyond conventional architectures into immersive and adaptive platforms. New paradigms such as 360-degree cylindrical panels, large-area direct-view 3D systems, colored e-paper, and flexible automotive dashboards illustrate how the industry is moving toward designs that combine structural adaptability with superior optical performance. These innovations demand coatings that are not only protective but also actively control reflection, scattering, and polarization, thereby enabling visual clarity under challenging environmental and mechanical conditions.

The discussion underscores the role of coatings in tailoring each application domain. For 360-degree advertising displays, anti-reflective and anti-smudge layers enhance visibility and durability in high-traffic indoor settings. Direct-view 3D displays require polarization-preserving coatings and retardation films to maintain image separation and depth perception over large viewing zones. Colored e-paper relies on scratch-resistant, anti-chemical coatings to preserve contrast and longevity in consumer and signage applications, while automotive dashboards integrate multilayer films for reflection suppression, scratch resistance, and moisture barriers to ensure readability and resilience in demanding environments. These examples illustrate how functional coatings are customized to meet application-specific optical and mechanical requirements.

The chapter concludes by emphasizing broader research directions. Nanostructured coatings, bio-inspired textures, and smart films with self-healing or adaptive properties are highlighted as enablers of long-term stability and multifunctionality. Scalable deposition techniques such as roll-to-roll, inkjet printing, and plasma-enhanced processes are identified as critical to transitioning laboratory concepts into mass-manufactured products. Collectively,

A. J. Olasoji and S. H. Im, *Functional Coating Layers for Surface Engineering of Electronic Displays*, SpringerBriefs in Applied Sciences and Technology, https://doi.org/10.1007/978-3-032-12843-0_7

these trends reveal that the convergence of advanced substrates with multifunctional coatings will define the future of display technologies, ensuring that next-generation systems achieve technical viability, commercial competitiveness, and user-centered performance.

The trajectory of commercial display technologies is rapidly shifting toward platforms that are not only visually immersive but also structurally adaptive and functionally integrated. Next-generation displays—such as 360-degree cylindrical or spherical panels, direct-view large-area 3D displays, high-resolution colored e-paper, and flexible automotive dashboard interfaces—are redefining the physical and optical boundaries of electronic visual systems. These innovations necessitate a parallel evolution in both substrate design and functional optical coatings to meet the growing demands for high mechanical compliance, environmental resilience, and superior optical clarity [1].

At the core of this transition is the requirement for advanced material systems that support unconventional form factors while maintaining—or even enhancing—display performance under variable conditions of use. This includes wide temperature and humidity ranges, repeated mechanical deformation (e.g., bending, stretching, or folding), and constant exposure to light, fingerprints, solvents, and atmospheric moisture. To address these multifaceted challenges, researchers are actively engineering substrates with customized modulus, thermal stability, and optical transmittance, often incorporating hybrid organic–inorganic composites and ultrathin nanolaminates.

Simultaneously, functional coating layers are being reimagined not merely as protective barriers, but as active optical interfaces capable of modulating light-matter interactions with subwavelength precision. Anti-reflective, anti-glare, and anti-smudge coatings are being augmented with additional capabilities such as polarization control, photonic filtering, self-healing, and adaptive transparency. Nanostructured coatings inspired by biomimetic designs (e.g., moth-eye surfaces or gyroid geometries) are increasingly employed to suppress surface reflections and improve image contrast across broad angular and spectral ranges. These coatings are also being designed for compatibility with scalable fabrication processes, including roll-to-roll, inkjet printing, plasma-enhanced chemical vapor deposition (PECVD), and atomic layer deposition (ALD), to enable cost-effective mass production.

Furthermore, the integration of multifunctional coatings with responsive or "smart" behavior—such as tunable refractive index, humidity-sensing, or anti-bacterial properties—opens new frontiers in healthcare, automotive, aerospace, and public information systems. These advancements are essential to ensuring that emerging display formats not only achieve technical feasibility but also deliver on long-term commercial viability, user engagement, and environmental sustainability. Ultimately, the synergy between advanced substrates and functional coatings will be instrumental in shaping the next paradigm of display technology: one that is immersive, adaptive, and seamlessly integrated into the fabric of daily life.

7.1 360-Degree Marketing and Advertisement Displays

The application of 360-degree display [2] is targeted for marketing and advertising of products and services by using advanced flexible active-matrix OLED, AMOLED display technology to ensure premium quality display definition. Deposition of optical coatings will enhance image display transmission while reducing light reflection especially in indoor installation setting where 360-degree displays are usually placed. Additionally, optical coatings will ensure protection from scratches, impacts, abrasions, and gashes to prevent damage to enhance device longevity since the displays will be located in direct contact with viewers [3, 4].

7.1.1 Purpose and Promise of 360-Degree Display Systems

360-degree displays are designed to deliver panoramic content visualization that can be seen clearly from all directions around a cylindrical or curved form factor. Unlike traditional flat screens, they eliminate viewing-angle limitations, making them ideal for high-traffic public environments where viewer orientation varies constantly. Their ability to attract attention from multiple directions simultaneously makes them particularly powerful for advertising, product promotion, and interactive installations [5].

To achieve such immersive experiences, these displays often rely on advanced OLED and AMOLED technologies, which are known for their high-contrast ratios, vivid color reproduction, deep blacks, and ultrathin form factors. Flexible variants of these displays are especially suited to the curvature required in 360-degree formats. Yet, this same flexibility and exposure also make them more vulnerable to environmental and mechanical stresses, necessitating protective and performance-enhancing solutions.

7.1.2 Challenges Faces 360-Degree Displays in Indoor Environments

360-degree displays are most commonly installed in indoor environments such as shopping malls, museums, luxury boutiques, airports, and high-end showrooms—settings where ambient lighting, close viewer proximity, and frequent physical contact present a distinct set of challenges for sustained display performance. One of the foremost issues is light reflection and glare caused by overhead lighting sources, including spotlights and fluorescent panels. These reflections can significantly reduce image clarity and legibility, especially when viewed from oblique angles or close range, which is typical for these installations. Even minimal surface glare can interfere with visual perception, detracting from the immersive viewing experience that

360-degree systems aim to provide. Additionally, mechanical wear and tear are major concerns due to continuous public interaction. Displays are highly susceptible to scratches, smudges, abrasions, and accidental impacts, all of which can degrade both the aesthetic appearance and the structural integrity of the screen over time. Compounding this, optical transmission losses may occur at the surface interface—even with high-definition OLED or AMOLED panels—primarily due to reflection, refraction, and scattering phenomena, which are further complicated by the curved geometries typical of 360-degree formats. To address these performance and durability bottlenecks, functional optical coatings are strategically engineered and either applied directly to the outer display surface or embedded within protective multilayer films. These coatings perform dual roles, providing essential optical enhancements such as anti-reflection and clarity retention, while also delivering mechanical protection against physical and environmental damage. As such, they are indispensable to ensuring the long-term functionality, visual clarity, and commercial effectiveness of 360-degree display systems in demanding indoor environments.

7.1.3 Enhancing 360-Degree Display Performance with Functional Optical Coatings

Flexible OLED and AMOLED panels have transformed the physical design of displays, enabling bending, folding, and curving into cylindrical or wraparound geometries. However, their flexibility also introduces mechanical limitations in terms of how much surface protection can be applied without compromising flexibility. Functional optical coatings must therefore be mechanically compliant—capable of maintaining their properties under bending stress without cracking or peeling. Advances in nanocomposite materials, sol–gel chemistry, and thin-film deposition (such as atomic layer deposition or plasma-enhanced techniques) now allow for coatings that are not only optically transparent and durable but also ultrathin and flexible [6].

In today's visually driven marketing landscape, the demand for innovative display technologies has intensified. Among the most promising advancements is the adoption of 360-degree display systems, which utilize advanced flexible active-matrix OLED and AMOLED panels. These displays offer seamless, immersive viewing experiences from all angles and are rapidly being deployed in high-end retail environments, exhibitions, transportation hubs, and other consumer-facing venues. However, to fully realize their commercial and technological potential, these systems must overcome several challenges related to optical clarity, environmental durability, and user interaction. This is where functional optical coatings play a pivotal role.

Optical coatings for display applications are typically categorized based on their specific functional roles, each contributing to enhanced visual performance, durability, and user experience [7]. One of the most essential categories is anti-reflective coatings, which are engineered to minimize surface reflections by optimizing the refractive index mismatch between the display surface and ambient air. This significantly improves light transmission and image visibility, especially in indoor environments where ceiling lights or spotlights can cause disruptive glare. Hard coatings, often formulated from silica-based or polymer-hybrid materials, are applied to resist mechanical abrasion and surface scratches while maintaining flexibility—an indispensable feature for interactive or touch-enabled displays frequently exposed to human contact. To preserve cleanliness and visual clarity, anti-smudge and oleophobic coatings are used to repel fingerprints, oils, and dust. These are particularly important for 360-degree displays installed at eye level or within reach, where maintaining surface hygiene is both a functional and aesthetic concern. In public and high-traffic settings, impact-resistant coatings play a critical role by protecting the display from accidental bumps or drops. These often incorporate impact-dampening films that absorb and disperse mechanical energy, preventing damage such as cracks or delamination in flexible display layers. Additionally, anti-UV and moisture-barrier coatings are vital for installations exposed to ambient sunlight through windows or high-humidity indoor environments. These coatings help preserve long-term device stability by shielding sensitive components—especially organic layers in OLED and AMOLED stacks [8]—from UV-induced degradation and moisture infiltration. Together, these diverse coating types form a comprehensive protection and enhancement system tailored for modern display technologies.

7.1.4   Advantages of 360-Degree Displays for Marketing and Commercial Use

From a marketing standpoint, the integration of functional optical coatings into 360-degree display systems delivers significant commercial advantages. Enhanced visual appeal is achieved through brighter, high-contrast, and glare-free images that more effectively capture consumer attention—particularly valuable in crowded advertising environments where visual impact is critical. Improved readability ensures that content remains crisp and legible across varying lighting conditions, allowing for consistent, uninterrupted messaging. Extended device lifespan is another key benefit, as coatings provide robust protection against scratches, abrasions, and environmental wear, ultimately reducing maintenance requirements and replacement costs. Additionally, boosted user confidence comes from displays that remain clean and unmarred despite frequent physical interaction, reinforcing a premium and high-quality user experience. Collectively, by enhancing durability, visual clarity, and overall display performance, optical coatings significantly strengthen the return on investment (ROI) for businesses deploying 360-degree display solutions in competitive, high-traffic settings.

7.1.5　Looking Forward

As 360-degree displays become more prevalent across industries—from retail marketing and museum exhibitions to corporate branding and smart environments—the role of functional optical coatings will only grow more central. Future directions may include self-healing coatings, smart coatings with adaptive light filtering, or multifunctional nanostructured films that combine anti-reflective, anti-fouling, and protective properties in a single layer.

Through continued research and integration of materials science, display engineering, and surface chemistry, optical coatings will remain vital in next-generation immersive display technologies—helping brands and organizations communicate more powerfully, durably, and beautifully.

7.2　Direct-View Large-Area 3D Displays

Originally, 3D display technology [9] was applied to 3D cinema usage [10]. Recently, the use of real 3D and direct-view images is gradually taking a leap into other display industry segments such as large-area displays which will require deposition of polarized optical coating that will not work well with regular 2D projected images. Hence, the large-area screen would require 3D display technology for the polarized display screens which are expensive and still very limited in function [11]. Therefore, optical coatings for multipurpose polarization of high-quality 3D displays are continuously under development for direct-view large-area 3D display technology.

7.2.1　The Shift from 3D Cinemas to Direct-View Applications

Historically, 3D imaging was predominantly confined to cinematic applications, where content was projected onto a large screen and viewed through polarized or anaglyph glasses. This setup, while effective in creating a depth illusion, was fundamentally reliant on projection systems and wearable visual aids. In recent years, however, advancements in panel-based 3D displays, including autostereoscopic and volumetric displays, have enabled a shift toward direct-view configurations—particularly in handheld devices, television panels, and now, increasingly, large-area commercial screens [12].

Unlike traditional 2D displays, these 3D systems require precise manipulation of light polarization and angle-dependent light guidance to ensure that different images reach each eye of the viewer to simulate depth. This demands an entirely new class of materials and coatings that can support multifunctional polarization while maintaining high brightness, resolution, and environmental durability.

7.2.2 *Limitations of Conventional 3D Coating Systems*

One of the main obstacles in scaling 3D displays [13–15] to large areas lies in the incompatibility of conventional optical coatings with 3D image projection. In typical 2D systems, coatings are optimized for general anti-reflection, light diffusion, and ambient contrast enhancement. However, 3D displays rely heavily on polarization-preserving optics—any disruption in polarization control leads to ghosting, image misalignment, or total loss of the 3D effect [16].

Furthermore, traditional coatings may introduce birefringence, scatter, or non-uniformity across large areas, especially when applied to curved or flexible substrates. These issues are magnified when scaling to meter-scale displays for advertising or public installations. Therefore, the need for specialized optical coatings—designed specifically to manage and preserve the polarization state of light in 3D display systems—is becoming increasingly urgent.

7.2.3 *Functional Optical Coatings for Direct-View Large-Area 3D Display Screens*

The field of display technology is undergoing a significant transformation as three-dimensional (3D) imaging moves beyond the realm of cinema into mainstream consumer electronics and commercial visualization systems. Among the most ambitious frontiers is the development of direct-view, large-area 3D display screens, which offer lifelike depth perception without the need for special glasses or headgear. These systems are poised to revolutionize industries such as digital signage, public information display, entertainment, and even medical imaging. However, the leap from concept to practical deployment is hindered by technical challenges—particularly in relation to polarization control and optical clarity. In this context, the role of functional optical coatings has become increasingly pivotal, offering potential solutions for optimizing light management, polarization efficiency, and display robustness.

To meet the stringent performance requirements of direct-view large-area 3D displays, several categories of advanced optical coatings are currently under active research or already adopted in commercial applications. One critical class is polarizing coatings, such as wire grid polarizers or nanostructured polarization-selective layers, which serve as integrated polarizers. These coatings selectively transmit one polarization state while blocking the orthogonal one and are typically laminated onto or embedded within the display surface [17]. They are particularly vital in glasses-free 3D display technologies, where controlling light polarization is essential for creating stereoscopic effects. Complementing these are retarder films and coatings, including quarter-wave and half-wave retarders, which convert light between different polarization states—such as from linear to circular polarization. When engineered as thin-film retarders, these coatings can reduce the complexity and thickness of the overall display stack, making them highly advantageous for large-area integration. Additionally, multilayer interference coatings are being designed to manipulate both the phase and amplitude of reflected light. These coatings can be precisely tuned to fulfill multiple optical roles simultaneously, such as polarization control, anti-reflection, and spectral filtering, thereby improving both efficiency and functionality in a single layer. For displays situated in public or high-traffic environments, anti-glare and hard optical coatings—which are carefully formulated to preserve polarization states—are essential. These coatings reduce surface glare while offering mechanical protection against abrasion, scratches, and daily wear, all without degrading the polarization integrity critical to 3D image formation. To achieve this, low-birefringence materials are often selected, ensuring that the display maintains high optical clarity and depth perception.

7.2.4 Polarization-Sensitive Coatings: The Foundation of 3D Visual Fidelity

Polarized 3D display systems operate by precisely controlling the orientation and propagation of light waves to create stereoscopic depth perception. In many advanced autostereoscopic and parallax barrier displays [18], left-eye and right-eye images are multiplexed using distinct polarization states—either linear or circular—so that each eye receives only the intended image. This polarization-based image separation is critical for delivering a convincing 3D effect without the need for external glasses. To support this function, specialized functional optical coatings are required that fulfill several stringent criteria. First, the coatings must enable polarization selection and maintenance, ensuring that the polarization state of the emitted or reflected light is preserved with minimal depolarization or optical loss. This demands the use of low-birefringence materials, precisely engineered anisotropic layers, and accurate alignment during deposition to accommodate the desired polarization geometry. Second, the coatings must help achieve minimal optical crosstalk between the stereoscopic image channels. Crosstalk—where light meant for one eye partially leaks into the other channel—can blur image boundaries and degrade the 3D visual experience. To combat this, polarization-selective coatings act as optical filters, suppressing unintended polarization components and enhancing image separation and clarity. Lastly, angular stability across the viewing field is essential. Because 3D displays rely heavily on directional light control, these coatings must maintain their polarization performance across a broad range of viewing angles. This ensures that viewers perceive consistent depth and image quality whether they are viewing the screen head-on or from an oblique position. Without such angular robustness, the 3D effect may break down or produce artifacts, limiting the display's effectiveness in real-world applications.

7.2.5 *Integration Challenges in Large-Area Direct-View Systems*

Scaling polarization-sensitive coatings for use in large-area 3D display systems [19] presents a distinct set of integration and performance challenges that go beyond those encountered in smaller or conventional displays. One of the most pressing concerns is achieving uniform coating thickness across the entire surface. In polarization-dependent systems, even slight variations in film thickness can significantly alter the phase retardation or polarization alignment, leading to spatial inconsistencies in 3D perception, image distortion, or crosstalk. This makes precision in deposition techniques—such as vacuum evaporation, roll-to-roll coating, or nanoimprint lithography—absolutely essential. Another major challenge lies in ensuring thermal and mechanical stability. Polarization-selective coatings, particularly those used on flexible substrates, must endure mechanical stress during bending or lamination, and also withstand elevated processing temperatures without undergoing structural deformation, delamination, or optical property degradation. Environmental robustness is equally critical, especially for large 3D displays installed in public or semi-outdoor environments such as shopping centers, museums, or transportation terminals. These coatings must resist moisture ingress, UV radiation, thermal cycling, and abrasion, which can otherwise cause polarization drift, haze, or surface damage. Additionally, engineers must navigate inherent optical trade-offs. Enhancing polarization efficiency—such as achieving a high extinction ratio or precise phase retardation—can often result in reduced display brightness, increased optical reflectivity, or added haze. Therefore, the development of these coatings requires careful multi-parameter optimization, balancing polarization control, transparency, durability, and visual clarity to meet the stringent demands of real-world large-format 3D display applications.

7.2.6 *Future Outlook: Multifunctional Coatings for Next-Gen 3D Displays*

The evolution of functional coatings for large-area 3D displays is moving toward multifunctional, nanostructured, and adaptive materials. For instance, liquid crystal polymer coatings, photonic crystals, and nanoimprinted metasurfaces are being developed to offer dynamic polarization control, allowing real-time modulation of depth effects based on viewer position or content. Moreover, angle-insensitive polarization-preserving coatings are becoming a major research focus, aiming to overcome the limited viewing zones of many current systems. Combined with improvements in display backplanes and micro-optics, these coatings could enable truly immersive, high-resolution 3D experiences on a building-scale screen.

As the display industry ventures beyond two-dimensional visualization into immersive 3D experiences, functional optical coatings will be indispensable for enabling large-area, direct-view 3D screens. These coatings serve as the linchpin between light control, polarization integrity, and durability, allowing 3D content to be displayed with clarity and robustness across diverse application environments.

While challenges remain—particularly in terms of scaling, cost, and integration—ongoing advances in materials science, nanofabrication, and thin-film engineering are poised to unlock the full potential of polarization-based 3D display systems. In doing so, they will bring us closer to a future where glasses-free, interactive, and lifelike 3D visualization becomes a ubiquitous part of our digital landscape.

7.3   Colored E-Paper Displays

The emergence of e-reader which typically uses electrophoretic display technology [20] produces texts and images that mimics real ink [21] on paper. Additionally, the application which does not need backlight has high-contrast display property to enhance readability and minimize strain on the eyes of viewers. Aside from application to e-book, e-paper can be diversified to other displays such as electronic billboard (signage), smartphones, smart wristwatches, photo albums, and restaurant menus. Optical coatings are required as topcoats for protection against scratch, abrasion, and impacts due to direct contact with fingers and sharp objects. These coatings are also expected to repel spreading of everyday chemicals that may come in contact with the display screen.

Colored e-paper extends the capabilities of traditional black-and-white electrophoretic displays by incorporating color pigments, color filters, or multielectrode architectures that enable the representation of full-color images. Unlike OLED or LCD technologies, which rely on emissive backlighting, e-paper displays are reflective and depend on ambient light, much like real paper. This makes them exceptionally energy-efficient and gentler on the eyes, as they do not emit harsh light [22].

As a result, colored e-paper displays are increasingly being considered for electronic price tags and store signage, outdoor and semi-outdoor information boards, smartphones and wearable devices, interactive menus and digital photo displays. Their low power consumption, sunlight readability, and printable-like aesthetic make them ideal for static or semi-static content in environments where conventional screens may fall short. However, as usage expands into more interactive and high-contact domains, protecting the integrity of the display surface becomes essential.

7.3.1 Functional Optical Coatings for Colored E-Paper Displays: Protecting the Paper of the Future

In recent years, the rise of e-readers and low power digital displays has ushered in a new era of screen technology, exemplified by the development and growing adoption of colored e-paper. These next-generation displays mimic the appearance of printed ink on real paper by utilizing electrophoretic display (EPD) principles—allowing for high-contrast, wide-angle visibility, and minimal eye strain, all without the need for a constant backlight. While monochrome e-paper has long been used in e-book readers, the expansion into color e-paper opens up a wide array of new applications, from electronic signage and smart wearables to interactive restaurant menus and photo albums.

Despite these advantages, colored e-paper displays face practical challenges, particularly in terms of surface durability, chemical exposure, and mechanical wear. These limitations can significantly impair user experience, especially in touch-based or high-contact settings. As a result, the integration of functional optical coatings has become increasingly important—not only to preserve optical clarity, but also to extend the operational lifespan and resistance of e-paper devices in real-world environments.

7.3.1.1 Surface Challenges in E-Paper Displays

While colored e-paper displays offer excellent visual performance and energy efficiency [23], they also present several surface-related vulnerabilities that must be addressed through advanced functional coatings. A primary concern is the risk of scratches and abrasions. E-paper devices, especially those used in consumer electronics or interactive retail signage, are frequently exposed to contact from fingernails, styluses, or even sharp objects. Given that these displays are typically constructed from flexible, polymer-based films, they are particularly prone to surface damage. Without adequate protective layers, such wear can lead to reduced legibility, diminished image quality, and shortened device lifespan. Another critical vulnerability is chemical sensitivity. In real-world use cases—such as restaurant menus, public kiosks, or transit signage—e-paper surfaces are likely to come into contact with food residues, skin oils, cleaning solvents, and other environmental contaminants. Without chemically resistant coatings, these substances can degrade surface layers, causing discoloration, ghosting, or permanent staining. Additionally, impact vulnerability remains a challenge. Although e-paper substrates offer some degree of flexibility, they are not impervious to mechanical stress such as dents, punctures, or delamination, especially from blunt impacts or repetitive physical force. To address this, coatings must be engineered to absorb and redistribute mechanical energy, thereby reducing the likelihood of structural damage. Finally, maintaining optical clarity and color integrity is essential. Protective coatings must not interfere with the display's inherent high-contrast reflectivity or color saturation. Any added layer should avoid introducing unwanted haze, glare, or color shift, and ideally, should enhance the optical performance without compromising display readability or vibrancy [22].

7.3.1.2   Role of Functional Optical Coatings in E-Paper Displays

To address the above limitations, specialized functional optical coatings are applied as top layers or laminated films over colored e-paper displays. These coatings serve a variety of purposes, from protection to optical optimization. Scratch-resistant and hard coatings introduce hard coatings, often made from silica composites or UV-curable polymers, form a durable surface shield that resists microabrasions from fingers, tools, or styluses. They are typically designed to maintain flexibility while enhancing surface hardness to 3H or higher on the pencil hardness scale. Chemical-resistant coatings prevent staining or degradation by oils, food, cleaning agents, or everyday chemicals, oleophobic and hydrophobic coatings are applied. These layers repel water- and oil-based substances, allowing the screen to stay clean and undamaged over time. Anti-glare and anti-Reflection coatings enhance readability. E-paper is naturally reflective, but external lighting—especially in glossy indoor settings—can still produce unwanted glare. Matte or nanostructured anti-glare coatings can diffuse incident light, improving readability without interfering with color contrast. Impact-resistant films useful for more rugged applications, such as outdoor signage or handheld devices, flexible, elastomeric coatings or laminated protective films can be employed to protect against shocks and pressure. These coatings absorb mechanical energy and prevent permanent deformation of the display surface. Anti-smudge and fingerprint-resistant coatings are also vital. Given the frequent touch interaction with e-paper devices, anti-smudge coatings will enhance the user experience by preventing visible fingerprint trails and making the surface easier to clean.

7.3.1.3   Design Considerations for Coating Integration in Colored
E-Paper Displays

When selecting or engineering functional coatings for colored e-paper displays, several critical design factors must be carefully considered to ensure both performance and durability. One of the foremost considerations is thickness and flexibility. Since e-paper technology is frequently implemented on flexible substrates, any applied coating must be ultrathin and mechanically compliant. Coatings that are too rigid or brittle risk cracking, peeling, or delaminating during bending or repeated mechanical stress, thereby compromising the structural integrity and visual quality of the display. Equally important is optical transparency. To preserve the inherent brightness, contrast, and color clarity of the e-paper, coatings must achieve high transmission rates—typically greater than 95%—with minimal light scattering or absorption. This is especially vital for colored e-paper, where subtle shifts in optical properties can noticeably affect image sharpness and saturation. Environmental stability is another key requirement. Coatings must provide long-term resistance to UV radiation, moisture, and thermal cycling, particularly for devices deployed in public-facing or semi-outdoor settings, such as transit signage, electronic menus, or outdoor retail displays. Without such protection, environmental exposure can lead to yellowing, delamination, or functional degradation of the display layers. Lastly, manufacturability and scalability must be considered. The selected coating formulations and processes should be compatible with high-throughput, large-area fabrication methods such as slot-die coating, spray deposition, gravure printing, or UV-curable techniques. This ensures that the coatings can be economically and uniformly applied over flexible substrates in roll-to-roll production settings, which is essential for the commercial viability of e-paper technologies.

7.3.1.4 Applications Benefiting from Optical Coating Integration in Colored E-Paper Displays

With the integration of functional optical coatings, colored e-paper displays become significantly more viable for a wide range of non-traditional display applications. For instance, interactive restaurant menus can benefit from scratch-resistant and waterproof coatings, enabling diners to browse or place orders directly on durable, wipe-clean digital surfaces. In smart wristwatches and smartphones, fingerprint-resistant coatings enhance tactile interaction while maintaining the premium matte finish characteristic of e-paper. For retail signage and electronic shelf labels, anti-glare and impact-resistant coatings improve legibility in high-traffic environments and help extend the functional lifespan of the display. Additionally, public transport information panels equipped with chemical- and abrasion-resistant coatings are better suited to withstand frequent human contact and prolonged environmental exposure, ensuring consistent performance and readability.

7.3.1.5 Future Prospects and Research Directions

Research into multifunctional hybrid coatings continues to accelerate. New developments include self-healing polymer coatings, anti-bacterial surfaces, and adaptive coatings that can respond to environmental stimuli like heat or pressure. These innovations promise to make colored e-paper displays not only more durable but also smarter and more interactive.

Moreover, biodegradable coatings for e-paper could open up sustainable applications in disposable or short-life electronics, such as event tickets, packaging, or smart labels. As colored e-paper technology advances into a broader range of interactive, flexible, and environmentally demanding applications, the role of functional optical coatings becomes more essential than ever. These coatings safeguard the visual quality, tactile experience, and structural integrity of the display—transforming e-paper from a niche technology into a robust platform for next-generation digital interfaces. By enhancing durability without compromising optical performance, functional coatings ensure that e-paper remains not only readable and reliable but also resilient enough for the dynamic demands of modern life.

7.4　Flexible Automobile Dashboard Displays

Traditional rigid electronic displays have long been integral to automobile dashboards, supporting a variety of digital interfaces such as speedometers, fuel gauges, and central touchscreens. However, as automotive design trends shift toward sleeker, more immersive, and ergonomically contoured interiors, rigid displays face limitations in conforming to the complex geometries of modern vehicle cabins. In response, flexible display technologies have emerged as a compelling alternative, offering seamless integration into curved or irregular surfaces without compromising functionality or aesthetics [24]. To ensure optimal performance in these environments, advanced optical coatings are indispensable. These include retardation films to enhance viewing angles, anti-reflective coatings to suppress unwanted glare, and protective layers that guard against scratches, chemical exposure, moisture, and mechanical abrasion. Together, these coatings play a critical role in enabling durable, high-clarity, and user-friendly flexible display systems tailored to next-generation automotive applications.

7.4.1　From Rigid Screens to Adaptive Design Interfaces

Traditional automotive b dashboards have long relied on rigid thin-film transistor liquid crystal display [25, 26] (TFT-LCD) screens for digital displays like speedometers, navigation, media interfaces, and fuel indicators. While effective, these displays are confined to flat or minimally curved sections of the dashboard, thereby restricting the creative freedom of automotive designers and limiting the display's integration within the vehicle interior. With modern vehicles increasingly adopting fluid, sculptural interior designs, automakers are turning to flexible display panels to enable seamless integration along wraparound dashboards, door panels, center consoles, and even steering wheels. These flexible OLED-based displays can bend and conform without sacrificing resolution, allowing for futuristic dashboard concepts and improved driver-user interfaces [24]. Yet the transition to flexibility introduces new complexities—most notably the need for specialized coatings that do not compromise flexibility while still offering optical and mechanical enhancements.

7.4.2 Display Performance Challenges in Automotive Environments

Unlike displays in conventional consumer electronics, automotive display systems must operate reliably under far more demanding environmental and mechanical conditions. These include sustained exposure to high ambient light, such as direct sunlight streaming through windshields, which can severely reduce screen readability due to glare and surface reflection. Additionally, displays must maintain visual clarity across wide-viewing angles, ensuring that both the driver and front-seat passenger can comfortably interpret the information displayed. Temperature extremes pose another challenge: automotive interiors can plunge below freezing in winter and soar to extreme heat in summer, especially in parked vehicles. Furthermore, displays are frequently subjected to moisture, dust, and chemical exposure—ranging from humidity and rain to aggressive cleaning agents used in car washes or detailing processes. Mechanical stresses such as repeated touch interactions, vibrations during motion, and occasional impacts also threaten the structural and optical integrity of the display. Without adequate protection, these factors can lead to image distortion, delamination, surface degradation, or even display failure. To overcome these issues, functional optical coatings are engineered to deliver tailored performance enhancements. Retardation coatings help maintain consistent brightness and color by compensating for birefringence and preserving polarization states. Anti-reflective coatings mitigate glare, improving legibility under intense lighting conditions. Hard, moisture-resistant, and chemically durable layers serve as protective barriers that shield the display from physical and environmental wear. Collectively, these coatings ensure that flexible automotive displays remain both visually effective and mechanically resilient throughout prolonged, real-world operation.

7.4.3 Functional Optical Coatings for Flexible Automobile Dashboard Displays

The automotive industry is currently undergoing a significant transformation, not only in terms of powertrain electrification and autonomous driving capabilities, but also in the design and integration of interior displays. Among these innovations, the shift from rigid electronic dashboard screens to flexible display technologies is becoming increasingly important. These next-generation displays, often based on OLED or AMOLED panels, offer the mechanical adaptability needed to conform to the complex curves, contours, and asymmetrical geometries found in modern vehicle interiors.

However, flexible displays face critical functional challenges—ranging from light reflection and narrow viewing angles to surface durability and chemical resistance. This is where functional optical coatings play a central role. Acting as multifunctional top layers, these coatings not only enhance display clarity and viewer comfort but also protect sensitive surfaces from wear and environmental degradation. As such, optical coatings are instrumental in enabling safe, reliable, and aesthetically integrated flexible dashboards for the next generation of smart vehicles.

7.4.3.1   Key Functional Optical Coatings for Flexible Automotive Displays

The application of functional coatings in automotive dashboard displays involves a sophisticated multilayered strategy, with each layer designed to address specific optical, mechanical, and environmental challenges. One critical layer is the retarder coating, also known as birefringence compensators, which mitigates viewing-angle distortions inherent in flexible OLED and AMOLED displays. These distortions—manifesting as shifts in brightness and color tone at oblique angles—are corrected using quarter-wave or half-wave retardation films that manipulate the polarization state of emitted light, thereby ensuring uniform image quality for both driver and passenger, regardless of their viewing position. To combat ambient light reflection, particularly from direct sunlight, anti-reflection (AR) coatings are applied to minimize surface glare by reducing the refractive index mismatch at the air–display interface. These coatings, often designed as multilayer thin films or nanotextured surfaces, deliver high transmittance levels (exceeding 95%) and very low reflectance, thus preserving screen legibility under diverse lighting conditions. Meanwhile, anti-scratch and hard coatings are essential for protecting dashboard displays from frequent mechanical interaction, including touch input and incidental contact with objects such as car keys, watches, or jewelry. Given the inherent softness of flexible display substrates, hybrid silica-polymer coatings are used to elevate surface hardness (typically to 3H or higher on the pencil hardness scale) without compromising flexibility. In addition, moisture and chemical barrier coatings are indispensable for safeguarding sensitive display layers against humidity, temperature fluctuations, and exposure to automotive cleaning agents. These coatings, characterized by their low water vapor transmission rates (WVTR) and high chemical resistance, prevent degradation of organic layers and preserve structural integrity over time. Lastly, to maintain a pristine, premium appearance, anti-fingerprint and smudge-resistant coatings—often based on oleophobic and hydrophobic chemistries—are applied to repel oils, dust, and moisture. These not only enhance the tactile experience but also reduce maintenance, supporting the aesthetic and functional expectations of high-end automotive interiors. Collectively, this integrated coating architecture ensures that automotive dashboard displays meet rigorous optical and mechanical standards while remaining visually consistent, durable, and user-friendly under real-world driving conditions.

7.4.3.2 Coating Integration with Flexible Substrates for Automotive Displays

One of the key challenges in applying optical coatings to automotive displays is the need for compatibility with flexible substrates. Conventional coatings may crack or delaminate when subjected to bending or thermal cycling. Advanced techniques such as sol–gel processes, plasma-enhanced chemical vapor deposition (PECVD), roll-to-roll coating, and UV-curing are now being used to deposit ultrathin, flexible, and conformable coatings that can bend and stretch along with the underlying display. In particular, hybrid nanocomposite coatings offer a balance between mechanical robustness and optical clarity, making them ideal for large-area flexible dashboards.

Automotive leaders such as BMW, Mercedes-Benz, and Tesla are already showcasing curved, edge-to-edge dashboard displays that rely on flexible OLED panels [27] and optical coatings for seamless integration. These displays not only serve functional roles—like showing speed, navigation, and media—but also double as ambient design elements, enhancing the overall aesthetic of the vehicle interior. As electric vehicles (EVs) and autonomous cars gain momentum, the need for smarter, more integrated, and durable in-car displays will only grow. Functional optical coatings will be a critical part of this transformation.

7.4.4 Future Directions

Looking ahead, researchers and developers are exploring multifunctional coatings that combine anti-reflective, hard, and self-healing properties into a single layer. There is also growing interest in adaptive coatings—such as those that respond to light or temperature—to dynamically optimize display performance.

Furthermore, sustainable and low volatile organic compound coating materials (low-VOCs) are being prioritized to align with the automotive industry's environmental goals. As vehicles evolve into intelligent, connected environments, the need for flexible, high-performance dashboard displays becomes increasingly critical. However, without proper protection and enhancement, even the most advanced display technologies can fall short of practical automotive demands. Functional optical coatings address these challenges by providing optical optimization, mechanical durability, and environmental resilience—making flexible displays viable and reliable for long-term use in automotive interiors. Through the integration of retarders, anti-reflective films, hard coatings, and barrier layers, display manufacturers and automakers can deliver immersive, safe, driverless, and future-ready in-car experiences [28]. In this context, coatings are not mere finishing touches—they are foundational components enabling the future of automotive design and functionality.

References

1. D. Saddington, Optical Coatings for Imaging and Displays published by IntertechPira 19 Northbrook Dr Portland, Maine 04105 USA (2012)
2. S.M. Liu, C.F. Chen, K.C. Chou, The design and implementation of a low-cost 360-degree color LED display system. IEEE Trans. Consum. Electron. **57**(2), 289–296 (2011)
3. D.W. Mohammed, Mechanical and electro-mechanical reliability of transparent oxide thin films for flexible electronics applications (Doctoral dissertation, University of Birmingham) (2016)
4. N. Shanmugam, R. Pugazhendhi, R. Madurai Elavarasan, P. Kasiviswanathan, N. Das, Anti-reflective coating materials: a holistic review from PV perspective. Energies **13**(10), 2631 (2020)
5. M. Shiina, N. Hashimoto, Transparent 360-degree display for high-resolution naked-eye stereoscopic aerial images, in *SIGGRAPH Asia 2024 Posters* (2024), pp. 1–2
6. J. Shahitha Parveen, M. Thirumurugan, New generation polymer nanocomposites for flexible electronics, in *Advances in Flexible and Printed Electronics: Materials, Fabrication, and Applications* (IOP Publishing, Bristol, UK, 2023), p. 2-1
7. H. Kim, K.H. Kim, Y.C. Jeong, Direct coating of transparent and wear-resistant polysilsesquioxane on ultra-thin glass for flexible cover windows. Prog. Org. Coat. **187**, 108162 (2024)
8. S.M. Lee, J.H. Kwon, S. Kwon, K.C. Choi, A review of flexible OLEDs toward highly durable unusual displays. IEEE Trans. Electron Devices **64**(5), 1922–1931 (2017)
9. N.S. Holliman, N.A. Dodgson, G.E. Favalora, L. Pockett, Three-dimensional displays: a review and applications analysis. IEEE Trans. Broadcast. **57**(2), 362–371 (2011)
10. M. Starks, 3DTV and 3D movie technology (2018)
11. J.A. Castellano, D.E. Mentley, Large-screen display industry: market and technology trends for direct view and projection displays, in *Projection Displays II*, vol. 2650 (SPIE, 1996), pp. 2–8
12. P. Benzie, J. Watson, P. Surman, I. Rakkolainen, K. Hopf, H. Urey, V. Sainov, C. Von Kopylow, A survey of 3DTV displays: techniques and technologies. IEEE Trans. Circuits Syst. Video Technol. **17**(11), 1647–1658 (2007)
13. S. Pastoor, M. Wöpking, 3-D displays: a review of current technologies. Displays **17**(2), 100–110 (1997)
14. S.J. Watt, K. Akeley, A.R. Girshick, M.S. Banks, Achieving near-correct focus cues in a 3D display using multiple image planes, in *Human Vision and Electronic Imaging* (2005), pp. 393–401
15. H. Fan, Y. Zhou, H. Liang, J. Wang, P. Krebs, J. Zhou, Displaying a full high-definition, high-quality 3D image without glasses. SPIE Newsroom **4**, 1–3 (2014)
16. L. Bégin, D. DePaoli, B. Bouma, D.C. Côté, M. Villiger, High fidelity polarization-sensitive optical coherence tomography through deep learning, in *Polarized Light and Optical Angular Momentum for Biomedical Diagnostics 2022* (SPIE, 2022), p. PC119630F
17. X. Shang, J. Niu, H. Li, L. Li, H. Hu, C. Lu, L. Shi, Polarization-sensitive structural colors based on anisotropic silicon metasurfaces. Photonics **10**(4), 448 (2023)
18. V. Ezhov, Phase-polarization parallax barriers for an autostereo/stereo/monoscopic display with full-screen resolution at each operation mode. Appl. Opt. **54**(28), 8306–8312 (2015)
19. T. Ni, G.S. Schmidt, O.G. Staadt, M.A. Livingston, R. Ball, R. May, A survey of large high-resolution display technologies, techniques, and applications, in *IEEE Virtual Reality Conference (VR 2006)* (IEEE, 2006), pp. 223–236
20. B.R. Yang (ed.), *E-paper Displays* (Wiley, 2022)
21. C. Am Kim, M.J. Joung, S.D. Ahn, G.H. Kim, S.Y. Kang, I.K. You, J. Oh, H.J. Myoung, K.H. Baek, K.S. Suh, Microcapsules as an electronic ink to fabricate color electrophoretic displays. Synth. Met. **151**(3), 181–185 (2005)
22. J. Heikenfeld, P. Drzaic, J.S. Yeo, T. Koch, A critical review of the present and future prospects for electronic paper. J. Soc. Inform. Display **19**(2), 129–156 (2011)

23. P.F. Bai, R.A. Hayes, M. Jin, L. Shui, Z.C. Yi, L. Wang, X. Zhang, G. Zhou, Review of paper-like display technologies (invited review). Prog. Electromagnet. Res. **147**, 95–116 (2014)
24. B.S. Yoon, C. White, G. Wease, L. Honnappa, S.T. Tsai, X. Wang, T.U. Daim, Technology roadmap for automotive flexible display, in *Planning and Roadmapping Technological Innovations: Cases and Tools.* (Springer International Publishing, Cham, 2013), pp.159–175
25. A.R. Shashidhara, G. Deepak, B.S. Manjunath, A. Murthy, T.N. Ruckmongathan, Liquid crystal display for an automobile dashboard (2004)
26. F. Bellotti, A. De Gloria, A. Poggi, L. Andreone, S. Damiani, P. Knoll, Designing configurable automotive dashboards on liquid crystal displays. Cogn. Technol. Work **6**(4), 247–265 (2004)
27. K. Blankenbach, Advanced automotive display measurements: selected challenges and solutions. J. Soc. Inform. Display **26**(9), 517–525 (2018)
28. A. Singh, Evaluating user-friendly dashboards for driverless vehicles: evaluation of in-car infotainment in transition (Master's thesis, Purdue University) (2023)

Chapter 8
Challenges and Research Directions in Optical Coatings for Future Display Technologies: Focus on Adhesion, Durability, and Optical Uniformity

This chapter investigates the central challenges that confront optical coatings as display technologies advance toward flexible, stretchable, foldable, and even holographic formats. While coatings have always been integral to managing reflection, transmission, and surface durability, their role is becoming increasingly complex with the adoption of unconventional substrates and multifunctional requirements. The discussion identifies adhesion, durability, and optical uniformity as the three most pressing bottlenecks that must be resolved for next-generation displays to achieve reliable performance. These interrelated challenges are framed within the broader context of how coatings interact with substrates, environmental stressors, and the demand for scalable manufacturing.

Adhesion challenges are explored first, particularly in relation to low-energy polymeric substrates and ultra-thin nanostructured films. Weak bonding leads to delamination, cracking, and premature coating failure, necessitating advances in surface functionalization, chemical coupling agents, and interlayer engineering. Durability emerges as another major issue, where coatings must withstand mechanical deformation, temperature cycling, UV exposure, and chemical attack without losing optical transparency or mechanical integrity. Optical uniformity is equally critical, as inconsistencies in thickness, refractive index, or nanostructure distribution introduce visible distortions, haze, or polarization artifacts, undermining image quality in high-resolution systems. The chapter also highlights emerging research directions aimed at overcoming these obstacles. Strategies include nanomaterial-based coatings with adaptive mechanical and optical properties, computational modeling and simulation to optimize coating design, and rigorous environmental testing protocols to ensure long-term reliability. Sustainability is introduced as an additional dimension, with emphasis on eco-friendly deposition methods and recyclable material

103

A. J. Olasoji and S. H. Im, *Functional Coating Layers for Surface Engineering of Electronic Displays*, SpringerBriefs in Applied Sciences and Technology, https://doi.org/10.1007/978-3-032-12843-0_8

systems. Collectively, the chapter makes clear that future innovation in adhesion chemistry, durability engineering, and optical precision will determine how coatings enable the transition from conventional displays to multifunctional, adaptive platforms that can meet the expectations of next-generation electronic display users for premium satisfaction.

The exponential growth of the display industry over the past few decades has led to a transformation in the way visual content is produced, transmitted, and consumed. From traditional cathode-ray tube (CRT) monitors to modern organic light-emitting diode (OLED) and micro-LED panels, the expectations for display quality, interactivity, and form factor have continuously evolved. At the heart of this evolution lies a critical but often underappreciated component: optical coatings. Optical coatings are multifunctional thin films applied to display substrates and surfaces to manipulate light behavior and improve optical clarity, durability, energy efficiency, and user interactivity. As industry moves toward next-generation displays—such as flexible, stretchable, foldable, transparent, and even holographic displays—the performance requirements placed on optical coatings are becoming more stringent. These advanced applications necessitate coatings that are not only optically superior but also mechanically resilient, environmentally stable, and integrable with unconventional substrates like ultra-thin polymers, nanostructured surfaces, and biodegradable materials.

The major challenges in developing optical coatings tailored for future (next-generation) display technologies specifically focus on three interrelated aspects: adhesion, durability, and optical uniformity. This chapter will delve into emerging research directions and material innovations aimed at addressing these critical albatrosses [1, 2].

8.1 Adhesion Challenges in Optical Coatings for Future Displays

As display technologies evolve toward thinner, more flexible, [3–5] and irregularly shaped formats, ensuring reliable adhesion between optical coatings and their substrates becomes increasingly critical. Adhesion governs the mechanical integrity and long-term performance of multilayer coatings, particularly under conditions of bending, stretching, heat exposure, and chemical contact. In future-oriented displays—ranging from foldable smartphones and rollable tablets to curved automotive dashboards and wearable devices—the complexity of substrate materials such as polyimide, polyethylene terephthalate (PET), or glass composites introduces unique surface energy and chemical compatibility challenges. This section explores the adhesion mechanisms at play, the factors that compromise coating-substrate interfaces, and the cutting-edge strategies being researched to enhance interfacial bonding for reliable and durable optical coatings [6, 7].

8.1.1 Substrate Incompatibility and Surface Energy Mismatch

One of the foremost challenges in coating development for future displays is ensuring robust adhesion to diverse and often unconventional substrates. Traditional coatings were primarily developed for flat, rigid substrates such as glass or silicon. However, flexible and wearable displays employ substrates like PET, polyimide (PI), thermoplastic polyurethane (TPU), and even biodegradable polymers. These substrates have inherently low surface energy and poor chemical affinity for conventional coating materials. As a result, coatings may delaminate, blister, or fail under thermal or mechanical stress. The adhesion issue is exacerbated when dealing with multi-layered display architectures that include barrier layers, electrode stacks, and interface films.

8.1.2 Influence of Mechanical Deformation

In foldable and rollable displays, the substrates and coatings are subjected to repeated bending, stretching, and torsion. Coatings that adhere well under static conditions may still fail when subjected to dynamic mechanical loading. The delamination risk increases at the interfaces where there is a mismatch in modulus of elasticity between the coating and the substrate. Moreover, the fatigue behavior of adhesion under cyclic deformation is not well understood for many coating-substrate systems, necessitating novel approaches in materials selection and adhesion engineering.

8.1.3 Sensitivity to Environmental Conditions

Adhesion is also influenced by environmental factors such as humidity, UV exposure, and temperature cycling. For instance, hydrophilic coatings may absorb moisture at the interface, leading to hydrolysis and weakening of adhesion. Similarly, thermally induced expansion or contraction creates interfacial stresses that promote peeling and cracking, especially in outdoor or automotive applications.

8.2 Research Directions to Improve Adhesion in Next-Generation Display Coatings

Improving adhesion between optical coatings and substrates remains a cornerstone of reliable performance in next-generation display technologies [8]. As display architecture becomes increasingly complex and flexible, conventional adhesion methods often fall short—especially when applied to substrates with low surface energy, such as plastics, or when coatings are subjected to repetitive mechanical or environmental stress. This has prompted researchers to explore innovative approaches that go beyond physical anchoring, focusing instead on molecular-level surface engineering, interfacial chemistry, and smart adhesion systems that respond adaptively to strain or thermal cycling. In this section, we examine emerging material strategies, interface design principles, and advanced processing techniques aimed at overcoming persistent adhesion limitations in future optical coatings.

8.2.1 Surface Treatment and Chemical Priming

To improve adhesion of functional coatings on low-surface-energy substrates such as polymers or treated glasses, a variety of surface modification techniques are being actively investigated. One of the most effective methods is plasma treatment, where exposure to oxygen or argon plasma introduces polar functional groups—such as hydroxyl, carbonyl, or carboxyl moieties—onto otherwise inert surfaces. This modification increases surface energy and enhances both wettability and chemical bonding affinity, thereby facilitating stronger adhesion of subsequent coating layers. Similarly, UV-ozone treatment offers a non-contact, uniform surface activation process that generates reactive oxygen species capable of cleaving surface contaminants and forming hydroxyl groups, particularly on polymeric and glass substrates. This significantly improves the performance of silane-based coupling agents used in adhesion promotion. Chemical primers, especially organosilanes like aminopropyltriethoxysilane (APTES) and glycidyloxypropyltrimethoxysilane (GPTMS), are also widely employed to form covalent linkages between the substrate and functional coating layers. These silane molecules bond to surface hydroxyls while simultaneously presenting functional terminal groups that interact with the incoming coating material, enabling durable and stable interfacial adhesion. Collectively, these surface treatments form the foundation of robust interface engineering essential for long-term coating performance in flexible and rigid electronic display applications.

8.2.2 Interfacial Engineering

Advanced interlayers with gradient modulus or chemical affinity for both coating and substrate can bridge the interfacial mismatch. For instance, grafted block copolymers or functionalized nanolayers can provide a stress-buffering region while enabling chemical compatibility.

8.2.3 Self-assembling and Layer-By-Layer (LBL) Coatings

Layer-by-layer assembly offers a pathway to construct coatings with controlled interfacial properties. By alternately depositing positively and negatively charged polyelectrolytes, films with tunable adhesion and surface energy can be built up with nanoscale precision. The incorporation of adhesive peptides, biomimetic moieties, or anchoring groups (e.g., catechols inspired by mussel adhesive proteins) adds further robustness.

8.3 Durability Challenges in Next-Generation Display Coatings

Durability is a defining performance metric for optical coatings, particularly in the context of next-generation electronic displays that are expected to operate under increasingly demanding mechanical and environmental conditions. As consumer expectations push toward thinner, lighter, and more flexible devices [9], coatings must simultaneously offer long-term resistance to abrasion, thermal cycling, chemical exposure, and UV radiation—often on unconventional substrates like polymer films or flexible glass. The challenge is further amplified in outdoor, wearable, or automotive applications where displays are continuously exposed to fluctuating temperatures, moisture, and physical impact. This section explores the multifaceted nature of durability in display coatings, identifying key failure mechanisms and material limitations, and lays the groundwork for targeted research into reinforcing coating robustness across emerging device platforms.

8.3.1 Mechanical Robustness and Fatigue

As flexible and wearable displays become ubiquitous, optical coatings are increasingly expected to maintain performance under repeated bending, compression, and even stretching. Traditional hard coatings such as silicon dioxide or aluminum oxide, while optically excellent, are brittle and fail under strain. Crack initiation and propagation in these coatings can result in loss of function, optical haze, or device failure. Additionally, the development of microcracks under fatigue loading leads to cumulative damage, even when individual strain cycles remain below the fracture threshold. This is particularly problematic in rollable e-paper or stretchable medical displays.

8.3.2 Scratch and Abrasion Resistance

Displays are subject to frequent handling and potential contact with abrasive particles, styluses, and cleaning agents. Coatings must resist microabrasions, which not only degrade optical clarity but also act as initiation sites for chemical or moisture ingress. Conventional anti-scratch coatings add rigidity but may conflict with the flexibility required for future displays.

8.3.3 Environmental Stability

Durability under environmental exposure represents a critical challenge in the long-term performance of functional coatings for electronic displays. Over time, coatings are subject to photochemical degradation, primarily due to prolonged ultraviolet (UV) exposure, which can lead to discoloration, surface embrittlement, and a reduction in optical transmittance. Thermal cycling, commonly encountered in outdoor and automotive environments, introduces repetitive expansion and contraction across multilayer structures, often resulting in interfacial stress, microcracking, or delamination due to mismatched thermal expansion coefficients between the coating and the substrate. Additionally, humidity and water ingress pose a serious threat—especially in unsealed or partially encapsulated devices—by accelerating hydrolytic degradation and weakening interfacial adhesion. Chemical exposure, including contact with skin oils, alcohol-based cleaners, and industrial solvents, can further deteriorate coating integrity by inducing swelling, dissolution, or breakdown of sensitive organic layers. These effects are exacerbated in flexible and wearable display systems, which endure complex environmental stressors such as body heat, perspiration, mechanical abrasion, and repeated flexing or folding. The synergistic impact of these factors significantly challenges coating stability and necessitates the development of advanced formulations with enhanced UV resistance, thermal stability, water repellency, and mechanical robustness.

8.4 Research Directions to Improve Durability in Next-Generation Display Coatings

To meet the escalating durability demands of next-generation electronic displays—especially those deployed in flexible [10], wearable, automotive, and outdoor environments—current research is focused on the development of multifunctional and resilient coating systems. One key direction involves the incorporation of UV-stabilizing agents such as nano-TiO_2, ZnO, or CeO_2 (cerium oxide), which can absorb or scatter harmful ultraviolet radiation, thereby mitigating photochemical degradation and discoloration over time. Parallel efforts are aimed at enhancing mechanical durability through the design of hybrid coatings that combine organic polymer matrices with inorganic nanofillers (e.g., silica nanoparticles, graphene oxide, or nanoclay), resulting in coatings with improved scratch resistance, reduced microcracking, and superior toughness without compromising flexibility. Barrier properties are also being aggressively optimized; advanced encapsulation strategies are exploring ultra-low water vapor transmission rate (WVTR) coatings using multilayer architectures or atomic layer deposition (ALD) of metal oxides to block moisture and oxygen ingress. Furthermore, research into self-healing coatings, utilizing microencapsulated repair agents or dynamic covalent bonding networks, is gaining traction to combat wear and tear from repeated mechanical stress. In addition, chemical resistance can be improved through fluorinated or siloxane-modified coatings that offer low surface energy and repel oils, solvents, and corrosive agents. The integration of these strategies—guided by predictive computational modeling and validated through accelerated environmental testing—is paving the way for durable, high-performance coatings that can withstand the harsh operating conditions expected of next-generation display technologies.

8.5 Optical Uniformity Challenges in Future Display Coatings

Optical uniformity is a critical requirement for modern electronic displays, directly impacting image quality, visual comfort, and color fidelity. As display resolutions increase and viewing angles expand, even the slightest inconsistency in coating thickness, refractive index, or surface morphology can produce undesirable effects such as color non-uniformity, haze, and image distortion. These issues are particularly pronounced in large-area or curved displays, where uniform light management across the entire surface is essential for seamless performance. In flexible and freeform displays, maintaining optical homogeneity becomes even more complex due to mechanical deformation and substrate variability. This section investigates the physical and material origins of optical non-uniformity in coatings, outlining the challenges associated with deposition processes, surface interactions, and environmental influences, while setting the stage for the design of next-generation coatings with precise and stable optical characteristics [11, 12].

8.5.1 Light Scattering and Surface Roughness

One of the most critical requirements in display coatings is achieving uniform light transmission and reflection across the entire surface. Even minor variations in coating thickness or surface morphology can cause unwanted optical phenomena such as iridescence, color shifting, or localized brightness anomalies. In display systems that employ backlighting or emissive pixels (e.g., OLEDs), non-uniform coatings can distort image clarity.

Surface roughness—especially at the nanoscale—can cause light scattering, leading to haze and reduced contrast. For example, in anti-reflective coatings, precise thickness control down to sub-nanometer precision is required to maintain destructive interference conditions. Variability across large-area panels, such as those used in televisions or automotive dashboards, is a frequent manufacturing issue.

8.5.2 Multilayer Thickness Inconsistency

Advanced optical coatings often consist of multiple layers with distinct refractive indices to produce interference effects (e.g., high-reflection, anti-glare, or dichroic behavior). Maintaining exact thicknesses and compositional uniformity across these layers is critical. Any deviation leads to spectral shifts and color non-uniformity across the display. This challenge is further exacerbated when applying coatings to non-planar or flexible substrates, where tension and curvature cause differential deposition rates and film stress gradients.

8.5.3 Refractive Index Mismatch and Interfacial Effects

Refractive index mismatches between the coating and substrate or between multiple layers can result in interfacial reflections, ghost images, and Fresnel losses. In transparent displays or see-through AR applications, such reflections are visually disturbing and interfere with the intended augmented overlay. In addition, the refractive index can be affected by environmental factors (e.g., temperature, humidity), resulting in dynamic shifts in optical performance—particularly problematic for outdoor or vehicular displays where the operating conditions fluctuate drastically.

8.6   Advanced Material and Design Strategies to Improve Optical Uniformity

Achieving high optical uniformity in coatings across various display types requires more than just precise deposition—it demands a deliberate integration of advanced materials and intelligent design strategies. Innovations in nanostructured films, gradient-index materials, and conformal deposition techniques have opened new pathways for minimizing light scattering, reducing interface mismatch, and maintaining spectral consistency across large or curved surfaces. These advancements are particularly relevant in high-resolution and flexible displays, where uniform optical response under dynamic mechanical conditions is crucial. In this section, we explore how novel material systems and tailored coating architectures—such as atomic layer deposition, nanoparticle composites, and scalable printing methods—are being leveraged to create coatings that combine aesthetic consistency with robust optical performance.

8.6.1   Atomic Layer Deposition (ALD) for Conformal Coating

ALD is a vapor-phase deposition technique that allows atomic-level control over layer thickness. It enables the creation of conformal coatings over complex topographies, which is particularly advantageous for curved or folded displays. Because the deposition process is self-limiting, ALD can achieve uniformity over large surfaces and in high aspect-ratio structures. Materials such as Al_2O_3, ZnO, and HfO_2 can be deposited using ALD for dielectric coatings, offering excellent optical clarity and refractive index control. Additionally, ALD is often combined with plasma-enhanced chemical vapor deposition (PECVD) for hybrid processing to improve throughput and mechanical resilience.

8.6.2   Gradient Index and Nanocomposite Layers

Gradient refractive index (grin) in coatings can reduce abrupt transitions between layers and minimize interfacial reflection. In the context of optical coatings, it refers to materials or layers in which the refractive index changes gradually (rather than abruptly) across the thickness of the film. This smooth transition helps reduce interfacial reflections and improves optical performance by mimicking a more natural light path through the material. These can be engineered by continuously varying the material composition during deposition, or by incorporating nanoparticles with tunable volume fractions. For example, TiO_2/SiO_2 nanocomposite coatings allow fine-tuning of refractive indices between ~ 1.45 and ~ 2.4, enabling customizable multilayer stacks for specific spectral properties. Nanostructures such as moth-eye surfaces can also create effective gradient indices via sub-wavelength patterning, producing broadband anti-reflective behavior with minimal scattering.

8.6.3   Inkjet and Slot-Die Coating for Large-Area Uniformity

To achieve large-area uniformity on flexible substrates, non-vacuum printing methods like inkjet, slot-die, and gravure coating are gaining popularity. These techniques allow patterned deposition of coating materials with high spatial control and minimal waste. Formulation of coating inks with optimal viscosity, solvent systems, and rheology is crucial for controlling film spreading and drying dynamics. Evaporation-driven instabilities (e.g., coffee-ring effect) can be mitigated by adjusting surface tension gradients and drying rates through additives and thermal controls.

8.7   Nanomaterial-Based Coatings for Next-Generation Displays

The incorporation of nanomaterials into optical coatings represents a transformative approach to meeting the evolving demands of next-generation display technologies. Nanomaterials such as graphene, carbon nanotubes (CNTs), metallic nanowires, and 2D transition metal dichalcogenides offer exceptional optical, electrical, and mechanical properties that traditional coating materials cannot match. Their tunable characteristics—ranging from high transmittance and conductivity to flexibility and environmental resilience—make them especially valuable in applications involving foldable, wearable, or transparent displays. Additionally, nanostructured coatings can be engineered to manipulate light at sub-wavelength scales, enabling new functionalities such as polarization control, enhanced light extraction, and adaptive modulation. This section delves into the types, roles, and integration challenges of nanomaterials in optical coatings, highlighting how they are shaping the future of advanced display systems [13–15].

8.7.1   Graphene and 2D Materials

Graphene and its derivatives offer exceptional optical, electrical, and mechanical properties. Few-layer graphene is nearly transparent ($\sim$ 97.7% transmittance) and highly conductive, making it a candidate for transparent electrode coatings. When used in hybrid coatings, graphene can improve scratch resistance, mechanical flexibility, and barrier properties without compromising optical clarity. Beyond graphene, other 2D materials such as MoS_2, WS_2, and h-BN are being explored for optical modulation, IR filtering, and as buffer layers to improve adhesion and reduce interface stress.

8.7.2 Carbon Nanotubes (CNTs) and Nanowires

CNTs can be used to reinforce mechanical durability while maintaining light transmission. For example, single-walled CNTs (SWCNTs) embedded in polymer matrices provide anisotropic reinforcement that resists crack propagation. Metallic nanowires, such as silver nanowire networks, serve as transparent conductors in touch screens and can be overcoated with protective and refractive index-tuned coatings. Challenges with nanomaterial integration include agglomeration, light absorption at certain wavelengths, and compatibility with flexible substrates. Surface functionalization and dispersion strategies are actively being researched to overcome these hurdles.

8.7.3 Photonic and Metamaterial Coatings

Photonic crystal coatings and metasurfaces can manipulate light in ways traditional coatings cannot. These engineered structures, composed of periodic dielectric or metallic nanostructures, can exhibit negative refractive index, selective angular filtering, and ultra-high reflectance at specific wavelengths. Applications include smart privacy filters, directional reflectors, and holographic projectors. While promising, challenges remain in scalable manufacturing, environmental stability, and cost-effective integration into display systems.

8.8 Computational Modeling and Simulation for Coating Design

As display architecture grows more intricate and performance demands intensify, computational modeling and simulation have become indispensable tools in the design and optimization of optical coatings. These techniques enable researchers to predict how coatings will behave across a spectrum of wavelengths, viewing angles, substrate geometries, and environmental conditions—well before physical prototypes are fabricated. By simulating layer stacks, interference effects, refractive index gradients, and angular dependencies, modeling platforms such as TFCalc, COMSOL Multiphysics, and OptiLayer allow for rapid iteration and fine-tuning of design parameters. This is especially critical for complex multilayer systems in foldable, large-area, or transparent displays, where real-world testing alone would be prohibitively time-consuming and expensive. In this section, we examine the methodologies, algorithms, and applications of computational modeling in optical coating development, and highlight how simulation bridges the gap between theoretical design and practical implementation [16, 17].

8.8.1   Optical Modeling and Optimization

Advanced simulation tools such as the transfer matrix method (TMM), rigorous coupled-wave analysis (RCWA), and finite-difference time-domain (FDTD) allow for detailed modeling of how multilayer coatings interact with incident light. These methods are widely used in optical engineering to simulate reflection, transmission, absorption, and field distribution in complex thin-film structures. These tools help optimize parameters such as layer thickness, angle of incidence, polarization, and wavelength range. By integrating simulation results into machine learning models, researchers are developing predictive algorithms that can propose optimal layer stacks for given optical targets (e.g., broadband AR, wavelength-selective filters). Such inverse-design frameworks significantly reduce experimental iteration.

8.8.2   Mechanical Stress and Deformation Modeling

Finite element modeling (FEM) tools (e.g., ANSYS, COMSOL) are used to simulate coating-substrate interactions under mechanical deformation. Parameters such as Young's modulus, Poisson's ratio, and interfacial adhesion energy are used to predict crack initiation and delamination zones. Coupled mechanical-optical simulations can predict changes in transmittance or birefringence under strain—critical for ensuring performance stability in flexible displays.

8.9   Environmental Testing and Reliability Assessment

Ensuring the long-term performance and safety of optical coatings in display technologies requires rigorous environmental testing and reliability assessment protocols. As displays are increasingly integrated into portable, wearable, and outdoor systems, coatings are subjected to a diverse array of stressors—including temperature cycling, humidity exposure, UV radiation, mechanical abrasion, and chemical contact—that can degrade optical quality and mechanical integrity over time. The reliability of a coating is not solely determined by its initial performance but by its ability to sustain functionality under operational and environmental duress. Consequently, advanced testing regimes—such as accelerated aging, thermal shock testing, and surface wear analysis—are essential to predict real-world performance. Environmental testing and reliability assessment explores the standardized methodologies, instruments, and failure criteria used to assess coating durability, offering insight into how reliability testing informs material choice and design optimization in next-generation display engineering [18]. To contextualize the detailed reliability protocols, Table 8.1 presents major environmental and mechanical testing procedures relevant to optical coatings in display applications. This guidance table presents test conditions, standardized methods, and practical pass/fail targets [19–31] in a concise format. By aligning with real-world stress factors—such as abrasion, thermal shock, folding cycles, and chemical exposure—the table provides researchers and engineers with a benchmark reference for evaluating coating durability across portable, wearable, automotive, and large-area display platforms.

Table 8.1 Consolidated guidance for environmental and reliability testing of optical coatings in display applications [19–31]

Test type	Typical conditions/ Standards	Pass/Fail targets	Application relevance
Abrasion (Taber test)	500–1000 cycles, CS-10 wheels, 500 g load	< 5% transmittance loss; haze increase < 1%	General AR/AG coatings, public displays
Pencil hardness	ASTM D3363	$\geq$ 3H (portable devices), $\geq$ 4H (automotive/ industrial)	Scratch resistance in touchscreens
Nanoindentation/ Microhardness	Depth-controlled indentation	Elastic recovery > 70%	Thin-film mechanical durability
Adhesion (cross-cut/ tape test)	ISO 2409/ASTM D3359	5B (no peeling/ delamination)	All display coatings
Sweat/Skin oil resistance	ISO 105-E04, artificial sebum immersion	No discoloration, $\Delta E^* < 2$, < 5% transmission loss	Wearables, smartphones
Sunscreen/Chemical resistance	UV-filtered sunscreen, 24–48 h contact	No delamination, < 5% haze increase	Outdoor devices, AR/ VR headsets
Thermal shock cycling	− 40 °C to + 85 °C, 500 cycles	No cracks, peeling, or > 2% change in T	Automotive, outdoor signage
High humidity test	85 °C/85% RH, 1000 h (JEDEC JESD22-A101)	$\Delta E^* < 2$, < 5% loss in transmittance, adhesion intact	OLED/QD displays
UV exposure (ASTM G154)	1000 h UV-A/B	< 5% loss in T, no yellowing	Outdoor displays, signage
Folding/Bending cycles	10 k–100 k cycles at 5–10 mm radius	No delamination, no visible cracks, $\Delta E^* < 2$	Foldable OLED/PI substrates
Drop-ball impact	0.5–1 J impact	No cracking or spallation	Ruggedized devices

8.9.1   Accelerated Environmental Testing for Coating Reliability Assessment

Standardized accelerated life testing plays a crucial role in evaluating the long-term reliability of functional coatings by simulating extended environmental exposure within a compressed timeframe. These tests are designed to replicate the harsh conditions that coatings may encounter in real-world applications, enabling early identification of failure mechanisms. Thermal cycling—typically ranging from $-40\ °C$ to $+85\ °C$—subjects coatings to repeated expansion and contraction, revealing susceptibility to delamination or cracking due to thermal stress. UV exposure tests, such as those outlined in ASTM G154, simulate the photochemical aging effects caused by prolonged sunlight, helping assess resistance to yellowing, embrittlement, and optical degradation. High humidity testing (commonly performed at 85 °C and 85% relative humidity for 1000 h) evaluates a coating's resistance to moisture ingress and hydrolytic instability. In addition, salt fog and chemical resistance tests are conducted to assess coating durability against corrosive agents and routine chemical exposures, such as those from skin oils or cleaning fluids. Throughout these evaluations, performance metrics such as optical transmittance, adhesion strength, hardness, and surface morphology are monitored using techniques like spectroscopic analysis, mechanical testing, and microscopy. These standardized protocols are essential for qualifying coatings for demanding applications in automotive, wearable, and outdoor display technologies.

8.9.2   Wear and Impact Resistance Testing

Mechanical durability is validated through abrasion (Taber abrasion test), pencil hardness, nanoindentation, and drop-ball impact testing. Wear mechanisms are studied through scanning electron microscopy and profilometry to characterize microcrack formation, coating loss, and material transfer. For flexible displays, bend testing (e.g., 10,000+ folding cycles) and dynamic fatigue testing are employed to assess long-term coating stability.

8.10   Sustainability Considerations in Coating Development

Traditional solvent-based coatings often emit volatile organic compounds (VOCs) and generate hazardous waste. As environmental regulations tighten, there is a growing demand for eco-friendly coating solutions. Due to electronics and display industries pushing toward greater environmental responsibility, sustainability has emerged as a critical design consideration in the development of optical coatings. Conventional coating processes often rely on energy-intensive vacuum deposition systems, hazardous solvents, and non-biodegradable materials, raising concerns about their ecological impact and regulatory compliance. With growing pressure from governments, consumers, and environmental watchdogs, researchers and manufacturers are now focusing on eco-friendly alternatives—such as waterborne coatings, solvent-free formulations, recyclable substrates, and deposition techniques that minimize material waste. These sustainable strategies must not only reduce environmental footprint but also maintain or enhance the optical and mechanical performance of the coatings. Sustainability considerations in coating development examine the core challenges and innovations in creating green optical coatings, emphasizing material choices, manufacturing practices, lifecycle assessments, and the role of sustainability in future display development [32–34].

8.10.1 Water-Based and UV-Curable Coatings

Water-based dispersions offer a lower environmental footprint and are increasingly viable for printable or sprayable coatings. UV-curable coatings, which polymerize upon exposure to UV light, eliminate the need for high-temperature curing and reduce energy consumption.

8.10.2 Biodegradable and Recyclable Materials

Biodegradable coatings using natural polymers (e.g., cellulose derivatives, chitosan) are under exploration for e-paper and short-lifetime displays. Recyclability is being enabled by designing coatings that can be depolymerized or dissolved under controlled conditions for substrate recovery.

8.10.3 Green Manufacturing Practices

Eco-friendly and sustainable fabrication strategies are increasingly being integrated into coating fabrication, including closed-loop deposition systems, low-temperature processing with usage of non-toxic solvents, and environmentally friendly additive manufacturing. These approaches collectively aim to minimize material waste, reduce energy consumption, and limit the need for extensive post-processing.

8.11   Outlook and Emerging Research Frontiers

The rapid evolution of display technologies—driven by innovations in flexibility, transparency, energy efficiency, and user interactivity—continues to place new demands on the design and performance of optical coatings. Looking ahead, the development of coatings will increasingly depend on breakthroughs in materials science, nanophotonics, and adaptive surface engineering. Future directions point toward smart coatings that can dynamically respond to stimuli such as light, heat, or strain; ultra-thin metamaterial-based coatings for precise control of electromagnetic waves; and multifunctional systems that integrate optical, electrical, and mechanical roles into a single layer. Moreover, as manufacturing shifts toward additive and scalable methods like roll-to-roll processing and inkjet printing, research must align with fabrication realities without compromising performance. This section outlines the most promising frontiers in optical coating research, reflecting the interdisciplinary collaboration needed to meet the complex challenges of next-generation display systems.

The trajectory of display technology toward ultra-flexible, adaptive, and high-resolution formats places enormous pressure on supporting optical coatings to perform under a broad spectrum of operational and environmental conditions. As we've explored, challenges in adhesion, durability, and optical uniformity are central to the success of next-generation displays. Research is moving rapidly to address these issues through interfacial engineering, smart material integration, predictive modeling, and green chemistry. The use of nanomaterials, metasurfaces, and adaptive coatings will redefine the limits of what optical films can achieve—allowing not just passive protection, but active light modulation, self-repair, and environmental sensing. Ultimately, interdisciplinary collaboration between material scientists, chemists, physicists, and engineers will be the key to driving coating innovation in step with display advances. The future of electronic displays—whether foldable smartphones, wearable AR glasses, or transparent automotive HUDs—depends critically on the invisible yet indispensable layer of innovation found in optical coatings.

References

1. W.S. Wong, A. Salleo (eds.), *Flexible Electronics: Materials and Applications*, vol. 11. (Springer Science & Business Media, 2009)
2. S. Wagner, S.J. Fonash, T.N. Jackson, J.C. Sturm, Flexible display enabling technology, in *Cockpit Displays VIII: Displays for Defense Applications*, vol. 4362. (SPIE, 2001), pp. 226–244
3. Z. Zhao, K. Liu, Y. Liu, Y. Guo, Y. Liu, Intrinsically flexible displays: key materials and devices. Natl. Sci. Rev. **9**(6), nwac090 (2022)
4. I.C. Cheng, S. Wagner, Overview of flexible electronics technology, in *Flexible Electronics: Materials and Applications*. (Springer, US, Boston, MA, 2009), pp.1–28
5. I.J. Chung, I. Kang, Flexible display technology–opportunity and challenges to new business application. Mol. Cryst. Liq. Cryst. **507**(1), 1–17 (2009)

6. K.L. Mittal (ed.), *Adhesion Measurement of Films and Coatings* (CRC Press, 2014)
7. W.B.A.U.S. Petasch, E. Räuchle, M.B.A.U.S. Walker, P.B.A.U.S. Elsner, Improvement of the adhesion of low-energy polymers by a short-time plasma treatment. Surf. Coat. Technol. **74**, 682–688 (1995)
8. D.S. Liu, C.Y. Wu, Adhesion enhancement of hard coatings deposited on flexible plastic substrates using an interfacial buffer layer. J. Phys. D Appl. Phys. **43**(17), 175301 (2010)
9. J. Si, Z. Li, X. Ren, Stretchable, soft, wear-resistant, transparent and recoverable Pu-Pa film for foldable displays. Available at SSRN 4910325
10. D.Y. Kim, M.J. Kim, G. Sung, J.Y. Sun, Stretchable and reflective displays: materials, technologies and strategies. Nano Convergence **6**(1), 21 (2019)
11. H. Fujiwara, *Spectroscopic Ellipsometry: Principles and Applications* (Wiley, 2007)
12. M. Ohring, *Materials Science of Thin Films: Deposition and Structure* (Academic Press, 2002)
13. H.T. Chen, A.J. Taylor, N. Yu, A review of metasurfaces: physics and applications. Rep. Prog. Phys. **79**(7), 076401 (2016)
14. A.K. Geim, Graphene: status and prospects. Science **324**(5934), 1530–1534 (2009)
15. T. Sannicolo, M. Lagrange, A. Cabos, C. Celle, J.P. Simonato, D. Bellet, Metallic nanowire-based transparent electrodes for next generation flexible devices: a review. Small **12**(44), 6052–6075 (2016)
16. M. Born, E. Wolf, *Principles of Optics: Electromagnetic Theory of Propagation, Interference and Diffraction of Light* (Elsevier, 2013)
17. P. Lalanne, P. Chavel, Metalenses at visible wavelengths: past, present, perspectives. Laser Photonics Rev. **11**(3), 1600295 (2017)
18. J.I. Goldstein, D.E. Newbury, J.R. Michael, N.W. Ritchie, J.H.J. Scott, D.C. Joy, *Scanning Electron Microscopy and X-Ray Microanalysis* (Springer, 2017)
19. ASTM D1044-20, *Standard Test Method for Resistance of Transparent Plastics to Surface Abrasion* (ASTM International, 2020)
20. ASTM D3363-20, *Standard Test Method for Film Hardness by Pencil Test* (ASTM International, 2020)
21. W.C. Oliver, G.M. Pharr, Measurement of hardness and elastic modulus by instrumented indentation: advances in understanding and refinements to methodology. J. Mater. Res. **19**(1), 3–20 (2004)
22. ISO 2409:2020. *Paints and Varnishes—Cross-Cut Test* (International Organization for Standardization, 2020)
23. ASTM D3359-17, *Standard Test Methods for Rating Adhesion by Tape Test* (ASTM International, 2017)
24. ISO 105-E04:2013, *Textiles—Tests for Colour Fastness—Part E04: Colour Fastness to Perspiration* (International Organization for Standardization, 2013)
25. IEC TR 62899-250:2016, *Printed Electronics—Part 250: Environmental Durability Test Methods* (International Electrotechnical Commission, 2016)
26. IEC 60068-2-14:2009, *Environmental Testing—Part 2-14: Tests—Test N: Change of Temperature* (International Electrotechnical Commission, 2009)
27. JEDEC JESD22-A104D:2014, *Temperature Cycling* (JEDEC Solid State Technology Association, 2014)
28. JEDEC JESD22-A101C:2014, *Steady-State Temperature Humidity Bias Life Test* (JEDEC Solid State Technology Association, 2014)
29. ASTM G154-21, *Standard Practice for Operating Fluorescent Ultraviolet (UV) Lamp Apparatus for Exposure of Nonmetallic Materials* (ASTM International, 2021)
30. IEC 62715-6-1, *Flexible Display Devices—Mechanical Performance Testing Methods* (International Electrotechnical Commission, 2021)
31. IEC 62262:2002, *Degrees of Protection Provided by Enclosures for Electrical Equipment Against External Mechanical Impacts (IK Code)* (International Electrotechnical Commission, 2002)
32. D.D. Paul, Optical metamaterials: fundamentals and applications (2010)

33. F. Qiu, H. Xu, Y. Wang, J. Xu, D. Yang, Preparation, characterization and properties of UV-curable waterborne polyurethane acrylate/SiO$_2$ coating. J. Coat. Technol. Res. **9**(5), 503–514 (2012)
34. W. Cai, V. Shalaev, Negative-index metamaterials, in *Optical Metamaterials: Fundamentals and Applications* (Springer, New York, 2009), pp. 101–122

Chapter 9
Conclusion and Prospects

This brief has given review and succinct perspectives on continuous evolution and trends of major electronic display technologies that nurtured the development of display screens—from cathode-ray tubes to organic light-emitting diodes (OLEDs), flexible screens, and large-area immersive systems—which have been fundamentally shaped by advancements in surface engineering, particularly through the deployment of functional optical coatings. It presents in-depth examination of the complex interplay between light, materials, and human vision, with emphasis on how engineered coating systems have underpinned the progress of displays used in televisions, computer monitors, smartphones, wearables, and public signage. More than mere finishing layers, these coatings constitute a critical interface between the display's internal photonic architecture and the surrounding environment, serving as active agents in managing optical behavior, environmental stability, and user comfort.

Modern display systems must fulfill a multitude of performance requirements—high luminance, wide color gamut, minimal reflectance, strong ambient readability, viewing angle consistency, and long-term mechanical durability. Achieving these attributes is not possible through display hardware alone; it necessitates the integration of multifunctional coating technologies that are finely tuned at micro- and nanoscales. Anti-reflective coatings, for instance, reduce unwanted light reflections and enhance contrast ratios, while hydrophobic or oleophobic coatings mitigate fingerprint smudging and chemical degradation. Polarization-controlling films manage light orientation for high-resolution LCDs and 3D stereoscopic displays, and multilayer interference coatings are indispensable for color filtering and spectral tuning in OLED and micro-LED panels.

As displays extend beyond rigid flat panels to more dynamic, flexible, and shape-conforming formats, coating technologies must evolve to match the mechanical and optical challenges posed by these emerging platforms. Foldable displays, for example, require coatings that are not only optically clear but also mechanically stretchable, delamination-resistant, and fatigue-tolerant over thousands of bending cycles. Similarly, 360-degree displays and large-format wraparound displays impose

A. J. Olasoji and S. H. Im, *Functional Coating Layers for Surface Engineering of Electronic Displays*, SpringerBriefs in Applied Sciences and Technology, https://doi.org/10.1007/978-3-032-12843-0_9

unique requirements on coating uniformity across complex geometries, demanding solutions such as conformal deposition techniques and gradient-index materials. Future displays integrated into automotive dashboards or wearable devices will be exposed to mechanical stress, UV radiation, temperature fluctuations, and chemical interactions, necessitating multibarrier coatings with advanced protective and adaptive functionalities.

One of the most compelling frontiers lies in intelligent or responsive coatings—those that change their optical behavior in real time in response to environmental cues such as light intensity, temperature, or user interaction. Electrochromic films, thermochromic coatings, and photoresponsive nanostructures are under intense development for their potential to modulate brightness, reduce power consumption, and improve eye comfort. In parallel, nanomaterial-based coatings incorporating graphene, quantum dots, or metallic nanowires enable ultra-thin, high-transparency, conductive layers that support both visual fidelity and electronic functionality.

From a manufacturing standpoint, the shift toward scalable and cost-effective processes such as roll-to-roll printing, inkjet deposition, and atomic layer deposition (ALD) has opened new possibilities for high-throughput production of complex coating systems with nanoscale precision. These techniques are essential to meet the rising demand for large-area, flexible, and high-resolution displays in consumer electronics, automotive, medical, and industrial sectors. Furthermore, eco-conscious innovations are becoming integral to coating design, with efforts directed at reducing solvent emissions, material waste, and environmental toxicity by exploring waterborne systems, biodegradable polymers, and recyclable substrates.

Looking ahead, the convergence of material science, photonics, and surface chemistry will continue to drive the functionality of optical coatings beyond traditional roles. Emerging trends in metamaterials, bio-inspired nanostructures, and machine-learning-assisted coating design will lead to optical surfaces that are not only visually optimized but also adaptive, intelligent, and multifunctional. For instance, bio-inspired coatings mimicking moth-eye or lotus-leaf structures provide anti-reflective and self-cleaning properties that are ideal for outdoor and wearable displays. Similarly, AI-driven modeling of multilayer interference stacks can accelerate the design of coatings with complex spectral profiles optimized for specific display technologies.

Finally, functional optical coatings are not peripheral considerations—they are foundational to the performance, reliability, and user experience of current and next-generation electronic displays. As display technologies progress toward more immersive, flexible, and sustainable formats, the coatings that govern their light–surface interactions must keep pace through innovation in material design, deposition techniques, and performance optimization. Investment in coating research and development is not merely complementary to display innovation—it is imperative. The future of coatings is therefore, in many ways, the future of electronic displays.